KU-608-839

Noise

The Political Economy
of Music

Jacques Attali

Translation by Brian Massumi
Foreword by Fredric Jameson
Afterword by Susan McClary

Theory and History of Literature, Volume 16

University of Minnesota Press
Minneapolis / London

The University of Minnesota Press gratefully
acknowledges translation assistance provided for
this book by the French Ministry of Culture.

Twelfth printing 2014

Published by the University of Minnesota Press
111 Third Avenue South, Suite 290, Minneapolis, MN 55401-2520
Printed in the United States of America on acid-free paper
Translation of *Bruits: essai sur l'économie politique de la
musique* © 1977 by Presses Universitaires de France

Cover illustration: *Carnival's Quarrel with Lent* by
Pieter Brueghel the Elder; Kunsthistorisches Museum, Vienna

Library of Congress Cataloging in Publication Data

Attali, Jacques.
  Noise.

  (Theory and history of literature; v. 16)
  Translation of: Bruits.
  Bibliography: p.
  Includes index.
  1. Music and society. 2. Music – Economic aspects.
I. Title. II. Series.
ML3795.A913   1985   780'.07   84-28069
ISBN 978-0-8166-1286-4
ISBN 978-0-8166-1287-1 (pbk.)

Noise

Theory and History of Literature
Edited by Wlad Godzich and Jochen Schulte-Sasse

Volume 1. Tzvetan Todorov *Introduction to Poetics*

Volume 2. Hans Robert Jauss *Toward an Aesthetic of Reception*

Volume 3. Hans Robert Jauss
Aesthetic Experience and Literary Hermeneutics

Volume 4. Peter Bürger *Theory of the Avant-Garde*

Volume 5. Vladimir Propp
Theory and History of Folklore

Volume 6. Edited by Jonathan Arac, Wlad Godzich,
and Wallace Martin
The Yale Critics: Deconstruction in America

Volume 7. Paul de Man *Blindness and Insight:
Essays in the Rhetoric of Contemporary Criticism*
2nd ed., rev.

Volume 8. Mikhail Bakhtin *Problems of Dostoevsky's Poetics*

Volume 9. Erich Auerbach
Scenes from the Drama of European Literature

Volume 10. Jean-François Lyotard
The Postmodern Condition: A Report on Knowledge

Volume 11. Edited by John Fekete *The Structural Allegory:
Reconstructive Encounters with the New French Thought*

Volume 12. Ross Chambers *Story and Situation: Narrative Seduction
and the Power of Fiction*

Volume 13. Tzvetan Todorov *Mikhail Bakhtin:
The Dialogical Principle*

Volume 14. Georges Bataille *Visions of Excess:
Selected Writings, 1927–1939*

Volume 15. Peter Szondi *On Textual Understanding and Other Essays*

Volume 16. Jacques Attali *Noise*

Contents

Foreword by Fredric Jameson   vii

Chapter One   Listening   3

Chapter Two   Sacrificing   21

Chapter Three Representing   46

Chapter Four   Repeating   87

Chapter Five   Composing   133

Afterword: The Politics of Silence and Sound
by Susan McClary   149

Notes   161

Index   171

Foreword
Fredric Jameson

The present history of music, *Noise*, is first of all to be read in the context of a general revival of history, and of a renewed appetite for historiography, after a period in which "historicism" has been universally denounced (Althusser) and history and historical explanation generally stigmatized as the merely "diachronic" (Saussure) or as sheer mythic narrative (Lévi-Strauss). The richness of contemporary historiography, however, by no means betokens a return to simpler narrative history or chronicle. Rather, the newer work can be seen as the renewal of a whole series of attempts, beginning in the nineteenth century, to write something like a totalizing history of social life, from the "expressive causality" of German *Geistesgeschichte*, or of Hegel himself, to Spengler or even Auerbach—all the way to the "structural causalities" of a Foucault or of the *Annales* school.

Music, however, presents very special problems in this respect: for while it is by no means absolutely unrelated to other forms and levels of social life, it would seem to have the strongest affinities with that most abstract of all social realities, economics, with which it shares a peculiar ultimate object which is *number*. The paradox is immediately underscored by the fact that the author of *Noise* is a professional economist; meanwhile, the recurrent phenomenon of child prodigies in music and in mathematics alike perhaps also suggests the peculiarity of the numerical gift, which would seem to demand less practical experience of the world and of social life than does work in other fields.

Yet, to use a well-worn Marxian formula, economics is generally considered to be a science of the base or infrastructure, whereas music traditionally counts among the most rarefied, abstract, and specialized of all superstructural activities. To propose intelligible links between these two "levels" or types of cultural and intellectual phenomena would therefore seem to demand the production of a host of intermediary connections or "mediations" that are by no means obvious or evident.

Max Weber, in whose time such methodological issues began to arise and whose own work poses them with a still unequaled lucidity, told this story in terms of a properly Western harmonic music whose very emergence constitutes an interesting historical problem in its own right ("Why was harmonic music developed from the almost universal polyphony of folk music only in Europe and only in a particular period, while everywhere else the rationalization of music took a different path—usually indeed precisely the opposite one, that of development of intervals by divisions of distance (usually the fourth) rather than by harmonic division [the fifth]?").[1] He summarized the complex determinants of the process as follows:

> Thoroughly concrete characteristics of the external and internal situation of the Church in the West, the result of sociological influences and religious history, allowed a rationalism which was peculiar to Western monasticism to give rise to these musical problems which were essentially "technical" in character. On the other hand, the invention and rationalization of rhythmical dancing, the origin of the musical forms which developed into the sonata, resulted from certain modes of social life at the time of the Renaissance. Finally, the development of the piano, one of the most important technical elements in the development of modern music, and its spread among the bourgeoisie, had its roots in the specifically "indoor" character of Northern European civilization.[2]

Weber's great story, as is well known, the master narrative into which the content of virtually all the research he ever did was reorganized, is that of the emergence of rationalization—so that it is not surprising to find western music described as one of the peculiar, forced products of this strange new influence (which Weber liked to derive from the rational enclave of monastic life in the Middle Ages). Elsewhere in the same essay he speaks of the "material, technical, social and psychological conditions" for a new style or art or medium; and it is obvious in the passage quoted above that the play of "overdetermination" between these conditions is complex indeed: material influences include, for example, the whole history of technology (and in particular the invention and production of musical instruments). What Weber calls the "technical" factor surely involves script or *notation*, a matter that in music goes well beyond a simple transcription of sounds and whose categories (tones, keys, etc.) will themselves generate and direct musical innovation. Meanwhile, the social realm, through the space of performance itself, simultaneously forms the public for music and its players alike; whereas the "psychological" confronts us with the whole vexed question of content and of ideology, and indeed ultimately the very problem of *value* that Weber's own historical analyses are explicitly concerned to suspend or to bracket. Weber notes, for instance, the association of "chromatics" with "passion," but adds: "It was not in the artistic *urge to* expression, but in the technical *means of* expression, that the difference lay between this ancient music and the chromatics which the great musical experimenters of the Renaissance created in their turbulent rational quest for new discoveries, and therewith for the ability to give musical shape to 'passion.' "[3] In Weber's brief remarks, it would seem that the word *passion* is meant to name a specific and historically original new form of psychological experience, so that the vocation of the newer music to express it is not, for Weber, a sign of value, but simply an item or feature necessary to complete the historical description.

It might indeed be well to distinguish two versions of the problem that begins to come into view here: one is that of a musical semantics, that is, of a relationship between musical signifiers and historical, social, psychological signifieds; the other is that of aesthetic value proper. Of the first of these Theodor Adorno has said: "If we listen to Beethoven and do not hear anything of the revolutionary bourgeoisie—not the echo of its slogans, but rather the need to realize them, the cry for that totality in which reason and freedom are to have their warrant— we understand Beethoven no better than does one who cannot follow the purely musical content of his pieces."[4] This seems straightforward enough until Adorno added what was always the "guiding thread" of Frankfurt School aesthetics: "Music is not ideology pure and simple; it is ideological only insofar as it is false consciousness."[5] The seeming contradiction between these two positions can perhaps be adjusted by a Habermasian appreciation of the universal (that is, nonideological) content of bourgeois revolutionary ideology as such; Adorno will himself complicate the situation more interestingly by factoring in the arrival of an age of aesthetic autonomy: "If [Beethoven] is the musical prototype of the revolutionary bourgeoisie, he is at the same time the prototype of a music that has escaped from its social tutelage and is aesthetically fully autonomous, a servant no longer."[6]

Yet if the question of musical value seems quite unavoidable when this line of inquiry is prolonged, the dramatic reversal we associate with the Russian Formalists is always possible: content of that kind (a new kind of passion, a new universal revolutionary ideology and enthusiasm, etc.) is itself the result of formal innovation. It is because the music of a given period is able to express new kinds of content that this last begins to emerge—a position which, translated back into linguistics, would yield a peculiar version of the Sapir-Whorf hypothesis. It is because language happens historically and culturally to be expanded in certain ways that we are able to think (and speak) this or that new thought.

There is nonetheless a gap between this semantic question about the nature of musical content and the problem of aesthetic value itself—a gap that Weber's own work registers with implacable lucidity, but that most subsequent historicisms have been at pains to disguise or conceal. There is just as surely a fundamental difference in emphasis as there is a spiritual kinship between Weber's interpretation of musical evolution in terms of the process of rationalization and Spengler's program for the morphology of cultures:

> The forms of the arts linked themselves to the forms of war and state-policy. Deep relations were revealed between political and mathematical aspects of the same Culture, between religious and technical conceptions, between mathematics, music, and sculpture, between economics and cognition-forms. Clearly and unmistakably there appeared the fundamental dependence of the most modern physical and chemical theories on the mythological concepts of our German ancestors, the style-congruence of tragedy and power-technics and up-to-date finance, and the fact (bizarre at first but soon self-evident) that oil-painting perspective, printing, the credit system, long-range weapons, and contrapuntal music in one case, and the nude statue, the city-state, and coin-currency (discovered by the Greeks) in another were identical expressions of one and the same spiritual principle.[7]

Value here becomes relativized according to the familiar patterns of historicism; and the value of music is then revealed when a larger, totalizing historical reconstruction of this kind allows us to read it as a fundamental expression of this or that basic cultural type. (It is not of any particular significance in the present context that Spengler also includes an ideological evaluation of those cultural types, the Faustian temporal dynamism of the West—including music above all!—being clearly for him "superior" to the spatial and Apollonian mode of Greek culture.)

The Frankfurt School, and most notably Adorno himself, sought escape from this kind of relativism by appealing to a Hegelian conception of aesthetic or formal self-consciousness. The utopian principle of value for these writers lies in freedom itself and in the conception of music as "the enemy of fate." Yet Adorno's other principle of evaluation is that of technical mastery, in which the superiority of a Schoenberg over a Hindemith, say, or a Sibelius, lies in the former's will to draw the last objective consequences from the historical state in which he found his own raw materials. These two principles, however, are capable, at certain moments in history, of entering into contradiction with one another, and not least, for Adorno, in the supreme moment of the achievement of the twelve-tone system itself:

> This technique . . . approaches the ideal of mastery as domination, the infinity of which resides in the fact that nothing heteronomous remains which is not absorbed into the continuum of the technique. It is, however, the suppressing moment in the domination of nature,

which suddenly turns against subjective autonomy and freedom itself, in the name of which this domination found its fulfill-
ment.[8]

Schoenberg's "moment of truth" is therefore to have replicated the dynamic of
a repressive, bureaucratic, and technocratic social order so completely as to
offer something like an aesthetic portrait or mirror image of it. Weber himself
was unwilling to praise musical rationalization as coming to formal consciousness of the deeper laws of Western social evolution; what was for him merely
one historical determinant among others becomes for Adorno the very episte-
mological function of art itself.

Yet these models of what we may call a musical historicism are all strangely
retrospective; at best, they grasp an achieved work, such as that of Beethoven
or Schoenberg, as reflecting (and illuminating or revealing) the dynamic of the
social system with which it was contemporaneous. The theoretical question of
whether such cultural forms simply replicate and reproduce that dynamic—or on
the contrary, distance, estrange, and criticize it—is a relatively secondary question, which depends on the problematic itself, which remains, according to
Marxian tradition, that of the relations of base and superstructure. No matter
that Engels, in his important late letters on historical materialism, tried to insist
on a "reciprocal interaction" between the economic and the superstructure:
most often such superstructures have in one way or another been taken as reflecting or corresponding to the economic, or at best as lagging behind concrete
social development. So Beethoven most richly expresses bourgeois revolutionary ideology, but after that ideology's triumph (and simultaneous failure), and
in the mode of an interiorization of more objective, collective, ideological
values.

The originality of Jacques Attali's book then becomes clear: he is the first
to have drawn the other possible logical consequence of the "reciprocal interaction" model—namely, the possibility of a superstructure to *anticipate* historical
developments, to foreshadow new social formations in a prophetic and annunciatory way. The argument of *Noise* is that music, unique among the arts for reasons that are themselves overdetermined, has precisely this annunciatory vocation; and that the music of today stands both as a promise of a new, liberating
mode of production, and as the menace of a dystopian possibility which is that
mode of production's baleful mirror image.

There are social and historical reasons for the orientation toward the future
in Attali's thought, as well as for the bleak bias toward the retrospective in
thinkers like Adorno, whose social and historical pessimism is documented in
Negative Dialectics. Not only is there Attali's greater sympathy for contemporary music, including that whole area of popular and mass-cultural production
which Adorno notoriously stigmatized as "easy music" and "jazz": degraded
and schematic commodities mass-produced by the Culture Industry. We must

also take into account Attali's economic thinking, as a practicing economist and a close adviser to president Mitterrand and as a central figure in France's current socialist experiment. Attali is also a distinguished scholar, the author of a dozen books whose subjects range from political economy to euthanasia. As varied as these books may seem, they have a common focus and a common problematic: the sense that something new is emerging all around us, a new economic order in which new forms of social relations can be discerned in the interstices of the old and of which new forms of cultural production can often give us the most precious symptoms, if not the prophetic annunciation.

Attali's varied and complex reflections thus rejoin, from a unique perspective (which is, given his political role, a unity of theory and practice in its own right), the now widespread attempts to characterize the passage from older forms of capitalism (the market stage, the monopoly stage) to a new form. This new form of capitalism, in which the media and multinational corporations play a major role, a shift on the technological level from the older modes of industrial production of the second Machine Revolution to the newer cybernetic, informational nuclear modes of some Third Machine Age. The theorists of this new "great transformation" range from anti-Marxists like Daniel Bell to Marxists like Ernest Mandel (whose work *Late Capitalism* remains the most elaborate and original Marxian model of some new third stage of capital).

For the most part, such efforts—including those of French poststructuralists like Jean Baudrillard and Jean-François Lyotard—have necessarily remained *historicist* in the inevitable positing of distinct *stages* of social development, whether the sequence of the latter is formulated in terms of evolutionary continuities or in those of breaks, ruptures, and cataclysmic mutations. Not all of Attali's own work escapes this temptation; but it is worth noting that one of his most recent syntheses, called *Les Trois Mondes* ("The Three Worlds"), seeks to "delinearize" the description of distinct social stages by modeling each in terms of distinct "world of representation" such that all three exist in our own time in a kind of synchronic overlap of the residual and emergent. The three worlds of Attali's title are not the more familiar geographical zones of the world system (with the "third world" positioned between the capitalist and the socialist countries), but rather are three distinct theoretical paradigms (ultimately generated by three distinct moments of history and of social organization). The first of these paradigms is that of *regulation*, conceived in mechanical terms of determinism and reversibility—a theory ultimately linked to the classical market. The second is that of *production*, whose strong form is clearly classical Marxism. The third paradigm Attali calls that of the *organization* of meanings and signs.

The positioning of Marxism as an older world paradigm clearly marks Attali's affinities with certain poststructuralisms and post-Marxisms, at the same time that it expresses the French Socialist party's complicated relationship to its

Marxian tradition. But unlike some of the more complacent celebrators of "post-industrial society" in the United States, Attali, as an economist in a socialist France that is in many ways the passive victim of a new (American) multinational order and of a worldwide economic crisis transcending the old nation-states, remains an essentially political thinker intent on discovering and theorizing the concrete possibilities of social transformation within the new system. His utopianism is thus materialistic and immanent, like that of Marx himself, who observed that the revolutionaries of the Paris Commune of 1871 "have no ideals to realize but to set free the elements of the new society with which old collapsing bourgeois society is itself pregnant."

Not the least challenging feature of Attali's thought lies in his tough-minded insistence on the ambiguity, or better still, the profound ambivalence, of the new social, economic, and organizational possibilities, which he often describes in terms of *autosurveillance*. From one perspective, autosurveillance marks the penetration of information technology within the body and the psyche of the individual subject: it implies a diffusion of computers on a generalized scale and a kind of passive replication of their programs by the individual, most visibly in the areas of education and medicine. Under autosurveillance, capital and the state no longer have to do anything to you, because you have learned to do it to yourself.

But "doing it to yourself" also implies knowing how to "do it for yourself," and the new technology is at least neutral to the degree that it could also, conceivably, be used for a collective political project of emancipation. Both dystopia and utopia are thus contained in the new forms as possibilities whose realization only political praxis can decide. For Attali's sense of social transformation is informed by the idea that radical change, the emergence of radically new social relations, is possible only as a result of the preexistence and the coincidence of three basic factors: "a new *technology* capable of reducing the costs of reorganization, *financial resources* (or an accumulation of new capital) available for the latter's utilization, and the existence of a *social group* with both the interest and the power to utilize such financial resources and to put the new technology to work."[9] The more properly socialist or political component of what might otherwise pass for a technolocratic vision of change is secured by the stress on this last factor—on the existence and praxis of a new *groupe moteur* or "subject of history."

The practical political and economic issues raised by Attali's other work are very far from being absent from *Noise*, despite its seemingly cultural focus. For one of the most stimulating features of this work is its insistence on grasping history and social life as a totality, in the way in which it offers us a model of the systematic interrelationship of the various levels of economics, technology, political forms, and culture proper. What is even more suggestive, however, in our current social, political, and historical confusion—in which the older strategies for radical social transformation and the older roads to some radically distinct utopian future have come to seem outmoded or unconvincing—is the prospective nature of his analysis, which stimulates us to search out the future in the present itself and to see the current situation not merely as a bundle of static and agonizing contradictions, but also as the place of emergence of new realities of which we are as yet only dimly aware. Jacques Attali's conception of music as prophetic of the emergent social, political, and economic forms of a radically different society can thus be an energizing one, whatever our judgment on the detail of his own analysis. In this work, we find, exceptionally in contemporary thought, a new model of the relations between culture and society that valorizes production in the present at the same time that it reinvigorates an enfeebled utopian thought.

University of California at Santa Cruz
February 1984

Noise

Chapter One
Listening

For twenty-five centuries, Western knowledge has tried to look upon the world. It has failed to understand that the world is not for the beholding. It is for hearing. It is not legible, but audible.

Our science has always desired to monitor, measure, abstract, and castrate meaning, forgetting that life is full of noise and that death alone is silent: work noise, noise of man, and noise of beast. Noise bought, sold, or prohibited. Nothing essential happens in the absence of noise.

Today, our sight has dimmed; it no longer sees our future, having constructed a present made of abstraction, nonsense, and silence. Now we must learn to judge a society more by its sounds, by its art, and by its festivals, than by its statistics. By listening to noise, we can better understand where the folly of men and their calculations is leading us, and what hopes it is still possible to have.

In these opening pages, I would like to summarize the essential themes of this book. The supporting argument will follow.

Among sounds, music as an autonomous production is a recent invention. Even as late as the eighteenth century, it was effectively submerged within a larger totality. Ambiguous and fragile, ostensibly secondary and of minor importance, it has invaded our world and daily life. Today, it is unavoidable, as if, in a world now devoid of meaning, a background noise were increasingly necessary to give people a sense of security. And today, wherever there is music, there is money. Looking only at the numbers, in certain countries more money is spent on music than on reading, drinking, or keeping clean. Music, an immaterial pleasure turned commodity, now heralds a society of the sign, of the immaterial up for sale, of the social relation unified in money.

It heralds, for it is *prophetic*. It has always been in its essence a herald of times to come. Thus, as we shall see, if it is true that *the political organization of the twentieth century is rooted in the political thought of the nineteenth, the latter is almost entirely present in embryonic form in the music of the eighteenth century.*

In the last twenty years, music has undergone yet another transformation. This mutation forecasts a change in social relations. Already, material production has been supplanted by the exchange of signs. Show business, the star system, and the hit parade signal a profound institutional and cultural colonization. Music makes mutations audible. It obliges us to invent categories and new dynamics to regenerate social theory, which today has become crystallized, entrapped, moribund.

Music, as a mirror of society, calls this truism to our attention: society is much more than economistic categories, Marxist or otherwise, would have us believe.

Music is more than an object of study: it is a way of perceiving the world. A tool of understanding. Today, no theorizing accomplished through language or mathematics can suffice any longer; it is incapable of accounting for what is essential in time—the qualitative and the fluid, threats and violence. In the face of the growing ambiguity of the signs being used and exchanged, the most well-established concepts are crumbling and every theory is wavering. The available representations of the economy, trapped within frameworks erected in the seventeenth century or, at latest, toward 1850, can neither predict, describe, nor even express what awaits us.

It is thus necessary to imagine radically new theoretical forms, in order to speak to new realities. Music, the organization of noise, is one such form. It reflects the manufacture of society; it constitutes the audible waveband of the vibrations and signs that make up society. *An instrument of understanding, it prompts us to decipher a sound form of knowledge.*

My intention here is thus not only to theorize *about* music, but to theorize *through* music. The result will be unusual and unacceptable conclusions about music and society, the past and the future. That is perhaps why music is so rarely listened to and why—as with every facet of social life for which the rules are breaking down (sexuality, the family, politics)—it is censored, people refuse to draw conclusions from it.

In the chapters that follow, music will be presented as originating in ritual murder, of which it is a simulacrum, a minor form of sacrifice heralding change. We will see that in that capacity it was an attribute of religious and political power, that it signified order, but also that it prefigured subversion. Then, after entering into commodity exchange, it participated in the growth and creation of capital and the spectacle. Fetishized as a commodity, music is illustrative of the evolution of our entire society: deritualize a social form, repress an activity of the body, specialize its practice, sell it as a spectacle, generalize its consumption, then see to it that it is stockpiled until it loses its meaning. Today, music heralds—regardless of what the property mode of capital will be—the establishment of a society of repetition in which nothing will happen anymore. But at the same time, it heralds the emergence of a formidable subversion, one leading to a radically new organization never yet theorized, of which self-management is but a distant echo.

In this respect, music is not innocent: unquantifiable and unproductive, a pure sign that is now *for sale*, it provides a rough sketch of the society under construction, a society in which the informal is mass produced and consumed, in which difference is artificially recreated in the multiplication of semi-identical objects. *No organized society can exist without structuring differences at its core. No market economy can develop without erasing those differences in mass production.* The self-destruction of capitalism lies in this contradiction, in the fact that music leads a deafening life: an instrument of differentiation, it has become a locus of repetition. It itself becomes undifferentiated, goes anonymous in the commodity, and hides behind the mask of stardom. It makes audible what is essential in the contradictions of the developed societies: *an anxiety-ridden quest for lost difference, following a logic from which difference is banished.*

Art bears the mark of its time. Does that mean that it is a clear image? A strategy for understanding? An instrument of struggle? In the codes that structure noise and its mutations we glimpse a new theoretical practice and reading: *establishing relations between the history of people and the dynamics of the economy on the one hand, and the history of the ordering of noise in codes on the other; predicting the evolution of one by the forms of the other; combining economics and aesthetics; demonstrating that music is prophetic and that social organization echoes it.*

This book is not an attempt at a multidisciplinary study, but rather *a call to theoretical indiscipline*, with an ear to sound matter as the herald of society. The risk of wandering off into poetics may appear great, since music has an essential metaphorical dimension: "For a genuine poet, metaphor is not a rhetorical figure but a vicarious image that he actually beholds in place of a concept."[1]

Yet music is a credible metaphor of the real. It is neither an autonomous activity nor an automatic indicator of the economic infrastructure. It is a herald, for change is inscribed in noise faster than it transforms society. Undoubtedly, music is a play of mirrors in which every activity is reflected, defined, recorded, and distorted. If we look at one mirror, we see only an image of another. But at times a complex mirror game yields a vision that is rich, because unexpected and prophetic. At times it yields nothing but the swirl of the void.

Mozart and Bach reflect the bourgeoisie's dream of harmony better than and prior to the whole of nineteenth-century political theory. There is in the operas of Cherubini a revolutionary zeal rarely attained in political debate. Janis Joplin, Bob Dylan, and Jimi Hendrix say more about the liberatory dream of the 1960s than any theory of crisis. The standardized products of today's variety shows, hit parades, and show business are pathetic and prophetic caricatures of future forms of the repressive channeling of desire.

The cardinal importance of music in announcing a vision of the world is nothing new. For Marx, music is the "mirror of reality"; for Nietzsche, the "expression of truth";[2] for Freud, a "text to decipher." It is all of that, for it is one of the sites where mutations first arise and where science is secreted: "If you close your eyes, you lose the power of abstraction" (Michel Serres). It is all of that, even if it is only a detour on the way to addressing man about the works of man, to hearing and making audible his alienation, to sensing the unacceptable immensity of his future silence and the wide expanse of his fallowed creativity. Listening to music is listening to all noise, realizing that its appropriation and control is a reflection of power, that it is essentially political.

The Sounds of Power

Noise and Politics

More than colors and forms, it is sounds and their arrangements that fashion societies. With noise is born disorder and its opposite: the world. With music is born power and its opposite: subversion. In noise can be read the codes of life, the relations among men. Clamor, Melody, Dissonance, Harmony; when it is fashioned by man with specific tools, when it invades man's time, when it becomes sound, noise is the source of purpose and power, of the dream—Music. It is at the heart of the progressive rationalization of aesthetics, and it is a refuge for residual irrationality; it is a means of power and a form of entertainment.

Everywhere codes analyze, mark, restrain, train, repress, and channel the primitive sounds of language, of the body, of tools, of objects, of the relations to self and others.

All music, any organization of sounds is then a tool for the creation or consolidation of a community, of a totality. It is what links a power center to its subjects, and thus, more generally, it is an attribute of power in all of its forms. Therefore, any theory of power today must include a theory of the localization of noise and its endowment with form. Among birds a tool for marking territorial boundaries, noise is inscribed from the start within the panoply of power. Equivalent to the articulation of a space, it indicates the limits of a territory and the way to make oneself heard within it, how to survive by drawing one's sustenance from it.[3] And since noise is the source of power, power has always listened to it with fascination. In an extraordinary and little known text, Leibnitz describes in minute detail the ideal political organization, the "Palace of Marvels," a harmonious machine within which all of the sciences of time and every tool of power are deployed.

These buildings will be constructed in such a way that the master of the house will be able to hear and see everything that is said and done without himself being perceived, by means of mirrors and pipes, which will be a most important thing for the State, and a kind of political confessional.[4]

Eavesdropping, censorship, recording, and surveillance are weapons of power. The technology of listening in on, ordering, transmitting, and recording noise is at the heart of this apparatus. The symbolism of the Frozen Words,[5] of the Tables of the Law, of recorded noise and eavesdropping—these are the dreams of political scientists and the fantasies of men in power: to listen, to memorize—this is the ability to interpret and control history, to manipulate the culture of a people, to channel its violence and hopes. Who among us is free of the feeling that this process, taken to an extreme, is turning the modern State into a gigantic, monopolizing noise emitter, and at the same time, a generalized eavesdropping device. Eavesdropping on what? In order to silence whom?

The answer, clear and implacable, is given by the theorists of totalitarianism. They have all explained, indistinctly, that it is necessary to ban subversive noise because it betokens demands for cultural autonomy, support for differences or marginality: a concern for maintaining tonalism, the primacy of melody, a distrust of new languages, codes, or instruments, a refusal of the abnormal—these characteristics are common to all regimes of that nature. They are direct translations of the political importance of cultural repression and noise control. For example, in the opinion of Zhdanov (according to a speech he gave in 1947 and never really disclaimed), music, an instrument of political pressure, must be tranquil, reassuring, and calm:

And, indeed, we are faced with a very acute, although outwardly concealed struggle between two trends in Soviet music. One trend represents the healthy, progressive principle in Soviet music, based upon recognition of the tremendous role of the classical heritage, and, in particular, the traditions of the Russian musical school, upon the combination of lofty idea content in music, its truthfulness and realism, with profound, organic ties with the people and their music and songs—all this combined with a high degree of professional mastery. The other trend is that of a formalism alien to Soviet art; it is marked by rejection of the classical heritage under the cover of apparent novelty, by rejection of popular music, by rejection of service to the people, all for the sake of catering to the highly individualistic emotions of a small group of aesthetes. . . . Two extremely important tasks now face Soviet composers. The chief task is to *develop* and perfect Soviet music. The second is to *protect* Soviet music from the infiltration of elements of bourgeois decadence. Let us not forget that the U.S.S.R. is now the guardian of universal musical culture, just as in all other respects it is the *mainstay of human civilization and culture against bourgeois decadence and decomposition of culture.* . . . Therefore, not only the musical, but also the political, ear of Soviet composers must be very keen. . . . Your task is to prove the superiority of Soviet music, *to create great Soviet music.*[6]

All of Zhdanov's remarks are strategic and military: music must be a bulwark against difference; for that, it must be powerful and protected.

We find the same concern, the same strategy and vocabulary, in National Socialist theorists. Stege, for example:

If Negro jazz is banned, if enemies of the people compose intellectual music that is soulless and heartless, and find no audience in Germany, these decisions are not arbitrary. . . . What would have happened if the aesthetic evolution of German music had followed the course it was taking in the postwar period? The people would have lost all contact with art. It would have been spiritually uprooted, all the more so since it would find little satisfaction in degenerate and intellectual music that is better suited to being read than heard. The gulf between the people and art would have become an unbridgeable abyss, the theater and concert halls would have gone empty, the composers working counter to the soul of the people would have been left with only themselves for an audience, assuming they were still able to understand their own wild fancies.[7]

The economic and political dynamics of the industrialized societies living under parliamentary democracy also lead power to invest art, and to invest in art, without necessarily theorizing its control, as is done under dictatorship. Everywhere we look, the monopolization of the broadcast of messages, the control of noise, and the institutionalization of the silence of others assure the durability of power. Here, this channelization takes on a new, less violent, and more subtle form: laws of the political economy take the place of censorship laws. Music and the musician essentially become either objects of consumption like everything else, recuperators of subversion, or meaningless noise.

Musical distribution techniques are today contributing to the establishment of a system of eavesdropping and social surveillance. Muzak, the American corporation that sells standardized music, presents itself as the "security system of the 1970s" because it permits use of musical distribution channels for the circulation of orders. The monologue of standardized, stereotyped music accompanies and hems in a daily life in which in reality no one has the right to speak any more. Except those among the exploited who can still use their music to shout their suffering, their dreams of the absolute and freedom. What is called music today is all too often only a disguise for the monologue of power. However, and this is the supreme irony of it all, never before have musicians tried so hard to communicate with their audience, and never before has that communication been so deceiving. Music now seems hardly more than a somewhat clumsy excuse for the self-glorification of musicians and the growth of a new industrial sector. Still, it is an activity that is essential for knowledge and social relations.

Science, Message and Time

"This remarkable absence of texts on music"[8] is tied to the impossibility of a general definition, to a fundamental ambiguity. "The science of the rational use of sounds, that is, those sounds organized as a scale"—that is how the *Littré*, at the end of the nineteenth century, defined music in order to reduce it to its harmonic dimension, to confuse it with a pure syntax. Michel Serres, on the contrary, points to the "extreme simplicity of the signals," "the message at its extreme, a ciphered mode of communicating universals" as a way of reminding us that beyond syntax there is meaning. But which meaning? Music is a "dialectical confrontation with the course of time."[9]

Science, message, and time—music is all of that simultaneously. It is, by its very presence, a mode of communication between man and his environment, a mode of social expression, and duration itself. It is therapeutic, purifying, enveloping, liberating; it is rooted in a comprehensive conception of knowledge about the body, in a pursuit of exorcism through noise and dance. But it is also past time to be produced, heard, and exchanged.

Thus it exhibits the three dimensions of all human works: joy for the creator, use-value for the listener, and exchange-value for the seller. In this seesaw between the various possible forms of human activity, music was, and still is, ubiquitous: "Art is everywhere, for artifice is at the heart of reality."[10]

Mirror

But even more than that, it is "the Dionysian mirror of the world" (Nietzsche).[11] "Person-to-person described in the language of things" (Pierre Schaeffer).

It is a mirror, because as a mode of immaterial production it relates to the structuring of theoretical paradigms, far ahead of concrete production. It is thus an immaterial recording surface for human works, the mark of something missing, a shred of utopia to decipher, information in negative, a collective *memory* allowing those who hear it to record their own personalized, specified, modeled meanings, affirmed in time with the beat—a collective memory of order and genealogies, the repository of the word and the social score.[12]

But it reflects a fluid reality. The only thing that primitive polyphony, classical counterpoint, tonal harmony, twelve-tone serial music, and electronic music have in common is the principle of giving form to noise in accordance with changing syntactic structures. The history of music is the "Odyssey of a wandering, the adventure of its absences."[13]

However, the historical and musicological tradition would still, even today, like to retain an evolutionary vision of music, according to which it is in turn "primitive," "classical," and "modern." This schema is obsolete in all of the human sciences, in which the search for an evolution structured in a linear fashion is illusory. Of course, one can perceive strong beats, and we will even see later on that every major social rupture has been preceded by an essential mutation in the codes of music, in its mode of audition, and in its economy. For example, in Europe, during three different periods with three different styles (the liturgical music of the tenth century, the polyphonic music of the sixteenth century, and the harmony of the eighteenth and twentieth centuries), music found expression within a single, stable code and had stable modes of economic organization; correlatively, these societies were very clearly dominated by a single ideology. In the intervening periods, times of disorder and disarray prepared the way for what was to follow. Similarly, it seems as though a fourth (and shorter) period was ushered in during the 1950s, with a coherent style forged in the furnace of black American music; it is characterized by stable production based on the tremendous demand generated by the youth of the nations with rapidly expanding economies, and on a new economic organization of distribution made possible by recording.

Like the cattle herd of the Nuer discussed by Girard,[14] a herd that is the mirror and double of the people, music runs parallel to human society, is structured like it, and changes when it does. It does not evolve in a linear fashion, but is caught up in the complexity and circularity of the movements of history.

This simultaneity of economic and musical evolution is everywhere present. We can, for example, toy with the idea that it is not by chance that the half-tone found acceptance during the Renaissance, at precisely the same time the merchant class was expanding; that it is not by coincidence that Russolo wrote his *Arte Dei Rumori* ("The Art of Noise") in 1913; that noise entered music and industry entered painting just before the outbursts and wars of the twentieth century, before the rise of social noise. Or again, that it is not by coincidence that the unrestricted use of large orchestras came at a time of enormous industrial growth; that with the disappearance of taboos there arose a music industry that takes the channelization of desire into commodities to such an extreme as to become a caricature; that rock and soul music emerged with the youth rebellion, only to dissolve in the cooptation of the young by light music programming; or finally, that the cautious and repressive form of musical production condoned today in countries with State-owned property designates "socialism" (if that is truly what it is) as simply the successor to capitalism, slightly more efficient and systematic in its normalization of men and its frantic quest for sterilized and monotonous perfection.

At a time when values are collapsing and commodities converse in place of people in an impoverished language (which in advertising is becoming increasingly musical), there is glaring evidence that the end of aesthetic codes is at hand. "The musical odyssey has come to a close, the graph is complete."[15]

Can we make the connections? Can we hear the crisis of society in the crisis of music? Can we understand music through its relations with money? Notwithstanding, the political economy of music is unique; only lately commodified, it soars in the immaterial. It is an economy without quantity. An aesthetics of repetition. That is why the political economy of music is not marginal, but premonitory. The noises of a society are in advance of its images and material conflicts.

Our music foretells our future. Let us lend it an ear.

Prophecy

Music is prophecy. Its styles and economic organization are ahead of the rest of society because it explores, much faster than material reality can, the entire range of possibilities in a given code. It makes audible the new world that will gradually become visible, that will impose itself and regulate the order of things; it is not only the image of things, but the transcending of the everyday, the herald of the future. For this reason musicians, even when officially recognized, are dangerous, disturbing, and subversive; for this reason it is impossible to separate their history from that of repression and surveillance.

Musician, priest, and officiant were in fact a single function among ancient peoples. Poet laureate of power, herald of freedom—the musician is at the same time within society, which protects, purchases, and finances him, and outside it, when he threatens it with his visions. Courtier and revolutionary: for those who care to hear the irony beneath the praise, his stage presence conceals a break. When he is reassuring, he alienates; when he is disturbing, he destroys; when he speaks too loudly, power silences him. Unless in doing so he is announcing the new clamor and glory of powers in the making.

A creator, he changes the world's reality. This is sometimes done consciously, as with Wagner, writing in 1848, the same year the *Communist Manifesto* was published:

> I will destroy the existing order of things, which parts this one mankind into hostile nations, into powerful and weak, privileged and outcast, rich and poor; for it makes unhappy men of all. I will destroy the order of things that turns millions into slaves of a few, and these

few into slaves of their own might, own riches. I will destroy this order of things, that cuts enjoyment off from labor.[16]

A superb modern rallying cry by a man who, after the barricades of Dresden, would adopt "the attitude of the rebel who betrayed the rebellion" (Adorno). Another example is Berlioz's call to insurrection:

> Music, today in the flush of youth, is emancipated, free: it does as it pleases. Many of the old rules are no longer binding: they were made by inattentive observers or ordinary spirits for other ordinary spirits. New needs of the spirit, the heart, and the sense of hearing are imposing new endeavors and, in some cases, even infractions of the old laws.

Rumblings of revolution. Sounds of competing powers. Clashing noises, of which the musician is the mysterious, strange, and ambiguous forerunner—after having been long emprisoned, a captive of power.

The Musician before Capital

The musician, like music, is ambiguous. He plays a double game. He is simultaneously *musicus* and *cantor*, reproducer and prophet. If an outcast, he sees society in a political light. If accepted, he is its historian, the reflection of its deepest values. He speaks of society and he speaks against it. This duality was already present before capital arrived to impose its own rules and prohibitions. The distinction between musician and nonmusician—which separates the group from the speech of the sorcerer—undoubtedly represents one of the very first divisions of labor, one of the very first social differentiations in the history of humanity, even predating the social hierarchy. Shaman, doctor, musician. He is one of society's first gazes upon itself; he is one of the first catalyzers of violence and myth. I will show later that the musician is an integral part of the sacrifice process, a channeler of violence, and that the primal identity *magic-music-sacrifice-rite* expresses the musician's position in the majority of civilizations: simultaneously *excluded* (relegated to a place near the bottom of the social hierarchy) and *superhuman* (the genius, the adored and deified star). Simultaneously a separator and an integrator.

In the civilizations of antiquity, the musician was often a slave, sometimes an untouchable. Even as late as the twentieth century, Islam prohibited believers from eating at the same table as a musician. In Persia, music was for a long time an activity restricted to prostitutes or, at least, considered shameful. But at the same time, the ancient religions produced a caste of musician-priests attached to the service of the temple, and mythology endowed musicians with supernatural and civilizing powers. Orpheus domesticated animals and transplanted trees; Amphion attracted fish; Arion built the walls of Thebes. The medicinal powers of music made musicians into therapists: Pythagoras and Empedocles cured the possessed, and Ismenias cured sciatica. David cured Saul's madness by playing the harp.

Despite the absence of an economic hierarchy in these societies, music was inscribed with precision into their systems of power. It is a reflection of the political hierarchy. So much so that many musicologists reduce the history of music to the history of the music of the princes.

Of course, in wealthy monarchies an orchestra has always been a display of power. In China, the musical code comprised five words: Palace, Deliberation, Horn, Manifestation, Wings.[17] Words of power. Words of subversion. What is more, in China the number and arrangement of the musicians indicated the position in the nobility of the lord who owned the orchestra: a square for the emperor, three rows for high dignitaries. The emperor authorized the forms of music that would assure good order within society, and prohibited those that might trouble the people. In Greece, even though there was no state supervision of music (with the exception of Sparta), and in Rome, where the emperors ensured their popularity by financing popular entertainment, music was essential to the workings of power. Throughout antiquity, then, we find the same concern for controlling music—the implicit or explicit channeler of violence, the regulator of society. Montesquieu understood this; he stated that for the Greeks music was a necessary pleasure—necessary for social pacification—and a mode of exchange—the only one compatible with good morals. He explicitly contrasted music to homosexuality and proclaimed their interchangeability:

> Why should music be pitched upon as preferable to any other entertainment? It is, because of all sensible pleasures, there is none that less corrupts the soul. We blush to read in Plutarch that the Thebans, in order to soften the manners of their youth, authorized by law a passion that ought to be proscribed by all nations.[18]

But a subversive strain of music has always managed to survive, subterranean and pursued, the inverse image of this political channelization: popular music, an instrument of the ecstatic cult, an outburst of uncensored violence. I am referring to the Dionysian rites in Greece and Rome, and to other cults originating in Asia Minor. Here, music is a locus of subversion, a transcendence of the body. At odds with the official religions and centers of power, these rites gathered marginals together in forest clearings and caves: women, slaves, expatriates. At times society tolerated them, or attempted to integrate them into the official religion; but at other times it brutally repressed them. There was a well-known incident in Rome that ended with hundreds receiving the death sentence.

Music, the quintessential mass activity, like the crowd, is simultaneously a threat and a necessary source of legitimacy; trying to channel it is a risk that every system of power must run.

Later, Charlemagne would forge the cultural and political unity of his kingdom by imposing the universal practice of the Gregorian chant, resorting to armed force to accomplish that end. In Milan, which remained faithful to the Ambrosian liturgy, hymnals were burned in the public square. A vagabond until the end of the thirteenth century, the musician subsequently became a domestic.

Vagabond

It took centuries for music to enter commodity exchange. Throughout the Middle Ages, the jongleur remained outside society; the Church condemned him, accusing him of paganism and magical practices. His itinerant life-style made him a highly unrespectable figure, akin to the vagabond or the highwayman.

The term *jongleur*, derived from the Latin *joculare* ("to entertain"), designated both musicians (instrumentalists and vocalists) and other entertainers (mimes, acrobats, buffoons, etc.). At the time, these functions were inseparable. The jongleur had no fixed employment; he moved from place to place, offering his services in private residences. He *was* music and the spectacle of the body. He alone created it, carried it with him, and completely organized its circulation within society.

The consumers of music belonged to every social class: peasants during the cyclic festivals and at weddings; artisans and journeymen at patron-saint celebrations; and at annual banquets, the bourgeoisie, nobles. A jongleur could very well play at a country wedding one night, and the next evening in the chateau, where he would eat and sleep with the servants. The same musical message made the rounds, and at each of these occasions the repertory was identical. Popular airs were performed at court; melodies composed in the palaces made it out to the villages and, in more or less modified form, became peasant songs. In the same way, the troubadours often wrote their poems to country airs.

Except for religious music, written music had not yet appeared. The jongleurs played from memory, an unvaried selection of melodies of their own composition, either very old peasant dances drawn from all over Europe and the Near East, or songs by noblemen or men of letters. If a melody was popular, numerous texts were based on it. All these styles functioned essentially within the same structures and were used interchangeably by the jongleurs, who effected a permanent circulation between popular music and court music.

In this precapitalist world in which music was an essential form of the social circulation of information, the jongleurs could be utilized for purposes of political propaganda. As an example, Richard the Lionhearted hired jongleurs to compose songs to his glory and to sing them in the public squares on market days. In wartime, jongleurs were often hired to compose songs against the enemy. Conversely, independent jongleurs composed songs about current events and satirical songs, and kings would forbid them to sing about certain delicate subjects, under threat of imprisonment.

We should, however, note two distinctive characteristics of the court musicians: first, certain highly learned and abstract texts of the troubadours were not sung in the villages. Second, only the courts had the means to hire, for major occasions, orchestras of jongleurs, composed of five or six musicians.

But with these two exceptions, music remained the same in the village, the marketplace, and the courts of the lords throughout the Middle Ages. The circulation of music was neither elitist nor monopolistic of creativity. The feudal world, with its polyphony, remained a world of circulation in which music in daily life was inseparable from lived time, in which it was active and not something to be watched.

In the fourteenth century, everything changed. On the one hand, church music became secularized and autonomous from the chant; it started to use an increasing number of instruments, incorporated melodies of popular and profane origin, and stopped relying exclusively on its Gregorian sources. On the other hand, the techniques of written and polyphonic music spread from court to court and distanced the courts from the people: nobles would buy musicians trained in church choirs and order them to play solemn songs to celebrate their victories, light songs for entertainment, orchestrated dances, etc. Musicians became professionals bound to a single master, *domestics*, producers of spectacles exclusively reserved for a minority.

Domestic

Within three centuries, from the fourteenth century to the sixteenth, the courts had banished the jongleurs, the voice of the people, and no longer listened to anything but scored music performed by salaried musicians. Power had taken hold, becoming hierarchical and distant. A shift in vocabulary confirms this mutation: the term *jongleur* was no longer used to designate a musician, but rather *ménestrel* ["minstrel"] or *ménestrier* [also "minstrel"], from the Latin *ministerialis*, "functionary." The musician was no longer a nomad. He had settled down, attached to a court, or was the resident of a town. When they were not domestics in the service of a lord, the minstrels organized themselves into guilds modeled after those of craftsmen or merchants, with a patron saint (St. Julian of the Minstrels), annual banquets, a retirement and disability fund, and dues set by municipal legislation. In exchange, they demanded and won a monopoly over marriages and ceremonies, shutting out the jongleurs, who were independent and often nonprofessional musicians. Since the courts had the means to finance resident musicians whom they held under exclusive control, the musicians acquired a new social position in Western society.

Until that time, the musician had been a free craftsman at one with the people and worked indifferently at popular festivals or at the court of the lord. Afterward, he would have to sell himself entirely and exclusively to a single social class.

Johann Joachim Quantz (1697-1773), who became flute-master to the Prussian king Frederick II after performing at town fairs, changing from jongleur to minstrel, gives a marvelous description of his experience of this mutation— from a time when music was a job like any other to a time when it was the occupation of specialists. From a time of the vagabond to a time of the domestic:

> My father was a blacksmith in the village. . . . In my ninth year, he began my training in the smithy's trade; even on his deathbed he declared that I had to continue in the trade. But . . . as soon as my father died, two brothers, one of whom was a tailor and the other a musician in the court of the town of Merseburg, offered to take me in and teach me their professions; I was free to choose which I preferred to adopt. From the age of eight, when I knew not a note of music, I insisted on accompanying my brother, who served as village musician in the peasant festivals, on a German bass viol, and this music, bad as it was, dominated my preference to such a degree that all I wanted was to be a musician. So I left for my apprenticeship in August of the year 1708, in Merseburg, under the above-mentioned Justus Quantz. . . . The first instrument I had to learn was the violin; I appear to have taken great pleasure in it and to have shown great skill. Then came the oboe and the trumpet. I worked especially hard on these three instruments during my three years of apprenticeship. As for the other instruments, like the cornet, the trombone, the hunting horn, the recorder, the bassoon, the German bass viol, the viola da gamba, and who knows how many others that a good musician must be able to play, I did not neglect them. It is true that, because of the number of different instruments one has in hand, one remains something of a bungler. However, with time one acquires that knowledge of their properties which is nearly indispensable for composers, especially those who write church music. The ducal chapel of Merseburg was not exactly rich at the time. We had to perform in church and at meals as well as at the court. When I finally finished my apprenticeship in December of the year 1713, I played several solos by Corelli and Telemann for the examination. My master excused me from three-quarters of a year of apprenticeship, but on the condition that I serve him a year longer in return for only half a journeyman's allowance. In March of 1718, the "Polish Chapel" was founded, which was to have twelve members. Since eleven members had already been chosen and

they needed an oboe player, I applied and, after an examination before the chapel master, Baron von Seyferitz, I was engaged into service. The annual salary was 150 taler, with free lodging in Poland. . . . I set about seriously studying the transverse flute, which I had also worked on: for I had no fear it would bring me animosity in the circle I was in. As a result of this new occupation, I began to think more seriously about composing. At that time there were not many pieces written specifically for the flute. . . . I left Dresden in December 1741, at which time I entered the king of Prussia's service. . . . [19]

Behind a mutation in the status of the musician, a rupture between two types of music.

The relations of reversibility between popular music and court music did not, however, end suddenly. Inspiration continued to circulate, to move between the classes. Since the capitalist system did not immediately replace the feudal system, the rupture between the two musical organizations was neither sudden nor total.

On the one hand, court musicians continued to draw from the popular repertory: they composed motets or masses based on songs from the streets, but they were unrecognizable in their polyphonic complexity. In the sixteenth century, collections of printed scores destined for customers in the courts—music's debut in the commercial world—offered orchestrations of popular dances and songs: "collections of songs both rustic and musical."

On the other hand, the jongleur did not disappear, and has not even to this day. Relegated to the villages, he suffered a decline in social status: he became the village minstrel, an ambulant musician who was often a beggar, or simply an amateur who knew how to sing or play the violin. But popular music no longer received much from music of the court, whose composers wrote works exclusively on demand, in particular for important events such as royal weddings, victory celebrations, coronations, funerals, or simply the visit of a foreign prince. One or two decades after its invention by the Florentine Camerata, opera became the most prominent sign of princely prestige. Every prince's marriage had its own original opera, the prologue of which would include an aria in praise of the sponsoring prince, a dedicatory epistle.

The musician, then, was from that day forward economically bound to a machine of power, political or commercial, which paid him a salary for creating what it needed to affirm its legitimacy. Like the notes of tonal music on the staff, he was cramped, chaneled. A domestic, his livelihood depended on the goodwill of the prince. The constraints on his work became imperative, immodest, similar to those a valet or cook was subjected to at the time. For example, the consistory of Arnstadt, on February 21, 1706, reproached the organist of its new church, Johann Sebastian Bach, for his private behavior:

Actum: The Organist of the New Church, Bach, is interrogated as to where he has lately been for so long and from whom he obtained leave to go.
Ille: He has been to Lübeck in order to comprehend one thing and another about his art, but had asked leave beforehand from the Superintendent.
Dominus Superintendens: He had asked only for four weeks, but had stayed about four times as long . . .
Nos: Reprove him for having hitherto made many curious variations in the chorale, and mingled many strange tones in it, and for the fact that the Congregation had been confused by it. In the future, if he wished to introduce a *tonus peregrinus*, he was told to hold it out, and not to turn too quickly to something else, or, as had hitherto been his habit, even play a *tonus contrarius*.[20]

A petty and impossible control to which the musician would be unceasingly subjected, even if in the bourgeois world of representation that control would be more subtle, more abstract than that which plagued Bach his entire life.

For all of that, however, the musician is not a mirror of the productive relations of his time. Gesualdo and Bach do not reflect a single ideological system any more than John Cage or the Tangerine Dream. They are, and remain, witnesses of the impossible imprisonment of the visionary by power, totalitarian or otherwise.

Understanding through Music

If we wish to elaborate a theory of the relations between music and money, we must first look at the existing theories of music. Disappointment. They are a succession of innumerable typologies and are never innocent. From Aristotle's three kinds of music—"ethical" (useful for education), "of action" (which influences even those who do not know how to perform it), and "cathartic" (the aim of which is to perturb and then appease)[21]—to Spengler's distinction between "Apollonian" music (modal, monodic, with an oral tradition) and "Faustian" music (tonal, polyphonic, with a written tradition), all we find are nonfunctional categories. Today, the frenzy with which musical theories, general surveys, encyclopedias, and typologies are elaborated and torn down crystallizes the spectacle of the past. They are nothing more than signs of the anxiety of an age confronted with the disappearance of a world, the dissolution of an aesthetic, and the slipping away of knowledge. They are no more than collections of classifications with no real significance, a final effort to preserve linear order for a material in which time takes on a new dimension, inaccessible to measurement. Roland Barthes is correct when he writes that "if we examine the current practice of music criticism, it is evident that the work (or its performance) is always translated with the poorest of linguistic categories: the adjective."[22]

So which path will lead us through the immense forest of noise with which history presents us? How should we try to understand what the economy has made of music and what economy music foreshadows?

Music is inscribed between noise and silence, in the space of the social codification it reveals. Every code of music is rooted in the ideologies and technologies of its age, and at the same time produces them. If it is deceptive to conceptualize a succession of musical codes corresponding to a succession of economic and political relations, it is because time traverses music and music gives meaning to time.

In this book, I would like to trace the political economy of music as a succession of *orders* (in other words, differences) done violence by *noises* (in other words, the calling into question of differences) that are *prophetic* because they create new orders, unstable and changing. The simultaneity of multiple codes, the variable overlappings between periods, styles, and forms, prohibits any attempt at a genealogy of music, a hierarchical archeology, or a precise ideological pinpointing of particular musicians. But it is possible to discern who among them are innovators and heralds of worlds in the making. For example, Bach alone explored almost the entire range of possibilities inherent in the tonal system, and more. In so doing, he heralded two centuries of industrial adventure. What must be constructed, then, is more like a map, a structure of interferences and dependencies between society and its music.

In this book, I will attempt to trace the history of their relations with the world of production, exchange, and desire; the slow degradation of use into exchange, of representation into repetition; and the prophecy, announced by today's music, of the potential for a new political and cultural order.

Briefly, we will see that it is possible to distinguish on our map three zones, three stages, three strategic usages of music by power.

In one of these zones, it seems that music is used and produced in the ritual in an attempt to make people *forget* the general violence; in another, it is employed to make people *believe* in the harmony of the world, that there is order in exchange and legitimacy in commercial power; and finally, there is one in which it serves to *silence*, by mass-producing a deafening, syncretic kind of music, and censoring all other human noises.

Make people Forget, make them Believe, Silence them. In all three cases, music is a tool of power: of ritual power when it is a question of making people forget the fear of violence; of representative power when it is a question of making them believe in order and harmony; and of bureaucratic power when it is a question of silencing those who oppose it. Thus music localizes and specifies power, because it marks and regiments the rare noises that cultures, in their normalization of behavior, see fit to authorize. Music accounts for them. It makes them audible.

When power wants to make people *forget*, music is ritual *sacrifice*, the scapegoat; when it wants them to *believe*, music is enactment, *representation*; when it wants to *silence* them, it is reproduced, normalized, *repetition*. Thus it heralds the subversion of both the existing code and the power in the making, well before the latter is in place.

Today, in embryonic form, beyond repetition, lies freedom: more than a new music, a fourth kind of musical practice. It heralds the arrival of new social relations. Music is becoming *composition*.

Representation against fear, repetition against harmony, composition against normality. It is this interplay of concepts that music invites us to enter, in its capacity as the herald of organizations and their overall political strategies—noise that destroys orders to structure a new order. A highly illuminating foundation for social analysis and a resurgence of inquiry about man.

For Fear, Clarity, Power, and Freedom correspond in their succession to the four stages Carlos Castaneda distinguishes in his mysterious description of the initiatory teachings of his master, the sorcerer Don Juan Mateus. This convergence is perhaps more than coincidental, if music is a means of understanding, like the unbalanced relation to ecstasy created by drugs. Is the sorcerer speaking of drugs when he explains that:

> When a man starts to learn, he is never clear about his objectives. His purpose is faulty; his intent is vague. He hopes for rewards that will never materialize, for he knows nothing of the hardships of learning. He slowly begins to learn—bit by bit at first, then in big chunks. And his thoughts soon clash. What he learns is never what he pictured or imagined, and so he begins to be afraid. Learning is never what one expects. Every step of learning is a new task, and the fear the man is experiencing begins to mount mercilessly, unyieldingly. . . . This is the time when a man has no more fears, no more impatient clarity of mind—a time when all his power is in check. . . . If a man . . . lives his fate through, he can then be called a man of knowledge, if only for the brief moment when he succeeds in fighting off his last, invincible enemy. That moment of clarity, power, and knowledge is enough.[23]

Don Juan's knowledge by peyote is reminiscent of the prophetic knowledge of the shaman, of the ritual function of the pharmakon. And of the interference between stages in the deployment of systems of music.

Music, like drugs, is intuition, a path to knowledge. A path? No—a battlefield.

Chapter Two
Sacrificing

Festival and Penitence, Violence and Harmony.[24] In an intense instability of powers, two processions, two camps, two lives, two relations to the World rumble and vie around a center of light and a well of darkness. Around them, the day-to-day labors of men, a strange round dance, boisterous child's play by the door to the church, and a cortege of penitents mark the significant figures of a secret dynamic—that of music and power.

For concealed behind the enactment of the conflict between religious order and its transgression in Festival lies every conceivable order. The poor wear masks and revel near a paltry tabernacle, while the rich observe Lent and flaunt their money by giving alms to the beggars arrayed outside the door to the church. In the Carnival parade, a musician, tragic and disquieting in the mask that disfigures him, stands beside men playing dice. Harmony and Dissonance. Order and Disorder. In this symbolic confrontation between joyous misery and austere power, between misfortune diverted into festival and wealth costumed in penitence, Brueghel not only gives us a vision of the world, he also makes it audible—perhaps for the first time in Western art. He makes audible a meditation on noise in human conflict, on the danger that festival will be crushed by a triumph of silence.

A meditation? A prophecy. Ambiguous and manifold. Open to all interpretations—and I would like to read it as a forecast of the path that music, trapped in the political economy, was to follow up to the present day.

Carnival's Quarrel with Lent is a battle between two fundamental political strategies, two antagonistic cultural and ideological organizations: Festival,

21

whose aim is to make everyone's misfortune tolerable through the derisory designation of a god to sacrifice; Austerity, whose aim is to make the alienation of everyday life bearable through the promise of eternity—the Scapegoat and Penitence. Noise and Silence.

Brueghel enacts this conflict in a space full of life: natural noises, noises of work and play, music, laughs, complaints, murmurs. Noises that today have virtually disappeared from our everyday life. An archeology of resonances but also of marginalities: for every character in the painting, with the exception of the bourgeois and the penitents, is marked by a physical deformity, woes of time. A cartography of deformities and a cartography of noises. A spatial pathology. A herald of regimentations in both the economic and religious orders. Brueghel saw the profound identity between noises and differences, between silence and anonymity. He announces the battle between the two fundamental types of sociality: the Norm and the Festival.

But he quite obviously had no delusions. In his time, the future was not in the festival, but in the norm; the Lent procession was marching triumphantly onward, supported by the bourgeoisie, women, and the Church. At the same time this tragic reversal took place—a reversal that would make religion an instrument of order, a prop for political power, the legitimator of suffering and regimenter of youth—Brueghel made audible six centuries of struggle to silence Festival in Ritual, Carnival in Sacrifice.

Three centuries before him, during which the Church decreed "that at saints' vigils, there shall not, in the churches, be any theater dances, indecent entertainment, gatherings of singers, or worldly songs, such as incite the souls of the listeners to sin" (Council of Avignon, 1209). During which it prohibited "granting assemblies of women, for the purpose of dancing and singing, permission to enter cemeteries or sacred places, regardless of considerations of dress," as well as prohibiting "nuns from heading processions, either within their own cloister or without, that circle churches and their chapels while singing and dancing, something which we cannot allow even secular women to do; for according to Saint Gregory, it is better, on Sunday, to toil and dig than to dance" (Council of Paris, 1212). During which it obliged "priests to prohibit, under penalty of excommunication, assemblies for dancing and singing from entering churches or cemeteries. . . . And if anyone performs dances before the churches of the saints, may they be subjected, if they repent, to three years' penance" (Council of Bayeux, beginning of the fourteenth century).

And three centuries after him, during which time the political economy would continue where the Church left off, pursuing this relegation to silence, domesticating musicians, and imposing its noises. Beginning in the seventeenth century, economic mechanisms break their silence and stop letting men speak. Production becomes noisy; the world of exchange monopolizes noise and the musician is inscribed in the world of money.

Today, Noise triumphs and reigns supreme over the sensibility of men. . . . Not only in the roaring atmosphere of major cities, but in the country too, which until yesterday was normally silent, the machine today has created such a variety and rivalry of noises that pure sound, in its exiguity and monotony, no longer arouses any feeling.[25]

Thus Brueghel's painting, in which music hardly figures, mysteriously brings out the essential dimension of its encounter with money and the political economy. First, music—a channelizer of violence, a creator of differences, a sublimation of noise, an attribute of power—creates in festival and ritual an ordering of the noises of the world. Then—heard, repeated, regimented, framed, and sold—it announces the installation of a new totalizing social order based on spectacle and exteriority.

Thus a colossal conflict pivots around a well, a point of catastrophe—a conflict between two social orders, two relations to power. Poverty is in both camps: on one side it is subjugated; on the other it transgresses. On one side, it is warm, luminous, and in solidarity. On the other, cold, dark, and solitary. Music is only in the light, near a tabernacle, the masked simulacrum of the pagan altar. There, it abuts dice and playing cards: music is the fragile order of ritual and prayer, an unstable order on the border of chance, harmony on the border of violence. Elsewhere, silence. Unless the penitents are accompanied by the *Lamentations*, or there is a hidden musician giving rhythm to the round dance.

This configuration of music, simultaneously visible and hidden, makes legible a cartography of the four essential forms of its political economy. While it visibly accompanies *Festival*, a simulacrum of the pagan sacrifice, and the Carnival parade of *Masks*, it is barely perceivable in the procession of the *Penitents* and the *Round Dance*.

Festival, Masks, Penitents, Round Dance. Four figures that pivot around a well and a stall, death and the commodity. The four possible statuses of music and the four forms a society can take. The conjunction, in one of the greatest works of Western painting, of this configuration of symbols and a modern conception of the dynamics of the political economy of music, is indeed surprising. It surprised me when I first thought that I read it in the painting, and it surprises me still. But Brueghel, in his meditation on the possible forms of noise, could not have failed to hear how they hinge on systems of power. He thus outlined everything it was possible to outline; he showed that we must not read into the painting a *meaning* of history, which is precluded by the circularity and shifting interpenetration of the figures; that we must use it instead to listen to music, which creates a ritual order, then is represented as a simulacrum of order, finally passes over to the side of Lent and is sold like fish, compulsory nourishment.

Even if one has no desire to hear his intuitions, it is impossible to avoid seeing this painting as a reminder that in the past the signification and role of music were conceived differently than they are today. Brueghel cries out that music, and all noises in general, are stakes in games of power. Their forms, sources, and roles have changed along with and by means of the changes in systems of power. Music trapped in the commodity is no longer ritualistic. Its code and original usage have been destroyed; with money, another code emerges, a simulacrum of the first and a foundation for new powers.

To establish a political economy of music, then, we must first rediscover this old code, decipher its meaning, in order to see how it was transformed by exchange into a use-value—a retrograde form, an impoverished memory of its ritual nature, the functional mode of the decline of the religious.

A sign: music has always been one. But it has been a deritualized, autonomous, commercial sign for too short a time for the study of its production and enjoyment to begin there. The political economy of a mode of human production cannot be elaborated without first inquiring into the social utility it had before being turned into a commodity, without first inquiring into the production of its usage. For example, the political economy of the chair presupposes an analysis of its usage, conducted prior to the study of its conditions of production. A priori, the usage of a chair is simple.[26] But the usage of music is obviously much more hermetic than that of the chair, even if there are more than enough pseudo-specialists around who still confine themselves to defining its usage by the pleasure derived from listening to it. In fact, it has no usage in itself, but rather a social meaning expressed in a code relating to the sound matter music fashions and the systems of power it serves. The concepts of political economy, designed to analyze the material world, are totally unsuited to this. They are blown apart by the study of the production and usage of signs.

To my way of thinking, and this is what I would like to establish in this chapter, the fundamental status of music must be deciphered through that of noise:

Noise is a weapon and music, primordially, is the formation, domestication, and ritualization of that weapon as a simulacrum of ritual murder.

In the succeeding chapters, I would like to demonstrate that music, after becoming a source of wealth, foreshadowed the destruction of codes; that is, I would like to show what capitalism—private or State—did to music and the noises of the body, and how it channeled, controlled, and repressed discourse and attempted to destroy meaning. Beyond that, I would like to ascertain what subversive renaissance is now under way.[27]

The Space of Music: From Sacrificial Code to Use-Value

Before exchange, we see that music fulfills a very precise function in social organization, according to a code I shall call *sacrificial*. Codification of this kind gives music a meaning, an operationality beyond its own syntax, because it inscribes music within the very power that produces society.

All music can be defined as noise given form according to a code (in other words, according to rules of arrangement and laws of succession, in a limited space, a space of sounds) that is theoretically knowable by the listener. Listening to music is to receive a message. Nevertheless, music cannot be equated with a language. Quite unlike the words of a language—which refer to a signified—music, though it has a precise operationality, never has a stable reference to a code of the linguistic type. It is not "a myth coded in sounds instead of words," but rather a "language without meaning."[28] It has neither meaning nor finality.

Thus when Saussure wants to fit music into the double structure of language, by distinguishing within it a signifier and a signified, he superimposes a semantic system on sounds: "We do not see what prevents a given idea from being associated with a succession of sounds"; Derrida implicitly does the same thing when he writes that "there is no music before language." This reasoning is not simply a theoretical hypothesis. For if it were accurate, music would not only necessarily be a transcribable, thus readable, discourse; but in addition all music separate from speech would have to be judged "degenerate" (Rousseau).[29]

To my mind, the origin of music should not be sought in linguistic communication. Of course, the drum and song have long been carriers of linguistic meaning. But there is no convincing theory of music as language. The attempts that have been made in that direction are no more than camouflages for the lamest kind of naturalism or the most mundane kind of pedantry. The musical message has no meaning, even if one artificially assigns a (necessarily rudimentary) signification to certain sounds, a move that is almost always associated with a hierarchical discourse.

In fact, the signification of music is far more complex. Although the value of a sound, like that of a phoneme, is determined by its relations with other sounds, it is, more than that, a relation embedded in a specific culture; the "meaning" of the musical message is expressed in a global fashion, in its operationality, and not in the juxtaposed signification of each sound element.

Use-Value and the Sacrificial Code

The operationality of music precedes its entry into the market economy, its transformation into use-value—the appropriation of "the materials of nature in a form adapted to [man's] needs."[30] Its primary function does not depend on the quantity of labor expended on it, but on its mysterious appositeness to a code of power, the way in which it participates in the crystallization of social organization in an order. I would like to show that this function is ritual in nature, in other words that music, prior to all commercial exchange, *creates political order because it is a minor form of sacrifice*. In the space of noise, *it symbolically signifies the channeling of violence and the imaginary, the ritualization of a murder substituted for the general violence, the affirmation that a society is possible if the imaginary of individuals is sublimated.*

This function of music gradually dissolves when the locus of music changes, when people begin to listen to it in silence and exchange it for money. There then emerges a battle for the purchase and sale of power, *a political economy.*

This hypothesis has many ramifications. It will be developed throughout the book; in this chapter, I will state it only in enough detail to give a glimpse of its theoretical range. The reader will perhaps have recognized it as an application in the domain of music of René Girard's broader discovery of the role of ritual sacrifice as a political channeler of and substitute for the general violence. It will be recalled that Girard demonstrated that the majority of ancient societies lived in terror of identity; this fear created a desire to imitate, it created rivalry, and thus an uncontrolled violence that spread like a plague—the "essential violence." By his account, in order to counteract this destruction of systems of social differences, all of these societies established powers, political or religious, whose role it was to block this dissemination of violence through the designation of a scapegoat. The sacrifice, real or symbolic, of the scapegoat polarized all of the potential violence, recreating differences, a hierarchy, an order, a stable society. Whence the peculiar status of the sacrificial victim, at once excluded and worshipped. Power and Submission. God and Nothingness.

In order to show that, before the commodity, music was a simulacrum of the sacrifice of the Scapegoat, and that it shared the same function, we must establish two things:

First, that *noise is violence*: it disturbs. To make noise is to interrupt a transmission, to disconnect, to kill. It is a simulacrum of murder.

Second, that *music is a channelization of noise*, and therefore a simulacrum of the sacrifice. It is thus a sublimation, an exacerbation of the imaginary, at the same time as the creation of social order and political integration.

The theory of noise's endowment with form, of its encoding, of the way in which it is experienced in primitive societies, should thus precede and accompany the study of the artifact that is the the musical work, the artifact whose endform is a sound-form that is the musical work.

The political economy of music should take as its point of departure the study of the material it highlights—noise—and its meaning at the time of the origin of mankind.

Noise, Simulacrum of Murder

A noise is a resonance that interferes with the audition of a message in the process of emission. A resonance is a set of simultaneous, pure sounds of determined frequency and differing intensity. Noise, then, does not exist in itself, but only in relation to the system within which it is inscribed: emitter, transmitter, receiver. Information theory uses the concept of noise (or rather, metonymy) in a more general way: noise is the term for a signal that interferes with the reception of a message by a receiver, even if the interfering signal itself has a meaning for that receiver. Long before it was given this theoretical expression, noise had always been experienced as destruction, disorder, dirt, pollution, an aggression against the code-structuring messages. In all cultures, it is associated with the idea of the weapon, blasphemy, plague. "Behold, I will bring evil upon this place, the which whosoever heareth, his ears shall tingle" (Jeremiah 19.3). "When the drums of the Resurrection sounded, they filled the ears with fear" (al-Din Runir, Divani, Shansi Tabriz).

In its biological reality, noise is a source of pain. Beyond a certain limit, it becomes an immaterial weapon of death. The ear, which transforms sound signals into electric impulses addressed to the brain, can be damaged, and even destroyed, when the frequency of a sound exceeds 20,000 hertz, or when its intensity exceeds 80 decibels. Diminished intellectual capacity, accelerated respiration and heartbeat, hypertension, slowed digestion, neurosis, altered diction: these are the consequences of excessive sound in the environment.

A weapon of death. It became that with the advent of industrial technology. But just as death is nothing more than an excess of life, noise has always been perceived as a source of exaltation, a kind of therapeutic drug capable of curing tarantula bites or, according to Boissier de Sauvages (in his *Nosologica methodica*), "fourteen forms of melancholy."

Music as a Simulacrum of Sacrifice

Since it is a threat of death, noise is a concern of power; when power founds its legitimacy on the fear it inspires, on its capacity to create social order, on its univocal monopoly of violence, it monopolizes noise. Thus in most cultures, the theme of noise, its audition and endowment with form, lies at the origin of the religious idea. Before the world there was Chaos, the void and background noise. In the Old Testament, man does not hear noise until after the original sin, and the first noises he hears are the footsteps of God.

Music, then, constitutes communication with this primordial, threatening noise—*prayer*. In addition, it has the explicit function of *reassuring*: the whole of traditional musicology analyzes music as the organization of controlled panic, the transformation of anxiety into joy, and of dissonance into harmony. Leibnitz writes:

> Great composers very often mix dissonance with harmonious chords to stimulate the hearer and to sting him, as it were, so that he becomes concerned about the outcome and is all the more pleased when everything is restored to order.[31]

We find the same process in operation all the way down to jazz.

The only piece of research that has been done on the relationship between violence and rock and roll—by Colin Fletcher—suggests that the music had a tendency to absorb violence, and to redirect violent energies into making music or into partisan but essentially nonviolent support for particular groups or singers.[32]

The game of music thus resembles the game of power: monopolize the right to violence; provoke anxiety and then provide a feeling of security; provoke disorder and then propose order; create a problem in order to solve it.

Music, then, rebounds in the field of sound like an echo of the sacrificial channelization of violence: dissonances are eliminated from it to keep noise from spreading. It mimics, in this way, in the space of sound, the ritualization of murder.

Music responds to the terror of noise, recreating differences between sounds and repressing the tragic dimension of lasting dissonance—just as sacrifice responds to the terror of violence. Music has been, from its origin, a simulacrum of the monopolization of the power to kill, a simulacrum of ritual murder. A necessary attribute of power, it has the same form power has: something emitted from the singular center of an imposed, purely syntactic discourse, a discourse capable of making its audience conscious of a commonality—but also of turning its audience against it.

The hypotheses of noise as murder and music as sacrifice are not easy to accept. They imply that music *functions* like sacrifice; that listening to noise is a little like being killed; that listening to music is to attend a ritual murder, with all the danger, guilt, but also reassurance that goes along with that; that applauding is a confirmation, after the channelization of the violence, that the spectators of the sacrifice could potentially resume practicing the essential violence.

The demonstration of this hypothesis in the context of the status of music in symbolic societies must go the route of an analysis of the relation between music and myth. Curiously, no general study exists on this quite fundamental question. And virtually nothing is known about the status of the music that was contemporary to myth.

Lévi-Strauss, for his part, has tried to show that music in our societies has become a substitute for myth. More exactly, he states that when the fugue—a form of composition identical to that of myth—was invented, one part of myth's heritage went into music, and the other into the novel. If so, it should then be possible to read the major musical forms from Monteverdi to Stravinsky as substitutes for the myths of symbolic societies. He deduces from this a classification system for musicians, distinguishing code musicians (Bach, Stravinsky, Webern) who comment on the rules of a musical discourse, from message musicians (Beethoven, Ravel, Schoenberg) who recount, from myth musicians (Wagner, Debussy, Berg) who code their message in narratives. Myth is then the production of a code by a message, of rules by a narrative.

But absent from his work is any explanation of the status of the music that was contemporary to myth. To my mind, music is not only a modern substitute for myth; it was present in myths in their time, revealing through them its primary operationality as a simulacrum of the ritual sacrifice and as an affirmation of the possibility of social order. The hypothesis of music as a simulacrum of sacrifice, which we have discussed up to now only in the context of a parallelism between the space of sound and the social space, is present, though in a subterranean way, in many myths. I will confine myself to three examples:

First, in the Chinese world, we find a great many very explicit formulations of these relations. For example, in Ssu-ma Ch'ien: "The sacrifices and music, the rites and the laws have a single aim; it is through them that the hearts of the people are united, and it is from them that the method of good government arises."[33]

In the Greek world, the myth of the sirens is quite clear: their song kills whoever hears it. The exception is Ulysses, who is offered as a simulacrum of the scapegoat, bound and incapable of commanding the obedience of his rowers, who are voluntarily deaf to protect themselves from the violence of the sound.

Finally, in the Germanic world, we find what is perhaps the best and most enlightening example of myths about music: the town invaded by rats, the attempts by the town officials to find a way to destroy them, the deal made with the piper, the drowning of the rats entranced by the music, the refusal of the officials to honor their agreement, and the piper's vengeance on the similarly entranced children—all of this makes the Pied Piper of Hamelin a cardinal myth. In fact, all of the elements of my hypothesis are present in it: music as the eliminator of violence, the usage of which is exchanged for money; the breaking of the contract; the reconversion of music into an instrument of violence against children. Money appears here as the sower of death, shattering the social unity that had been restored through music.

To my way of thinking, music appears in myth as *an affirmation that society is possible.* That is the essential thing. Its order simulates the social order, and its dissonances express marginalities. *The code of music simulates the accepted rules of society.* It is in this connection that the debate on the existence of a natural musical code and an objective, scientific, universal harmony takes on importance; we will return to this question later. If such a code did in fact exist, then it would be possible to deduce the existence of a natural order in politics and a general equilibrium in the economy.

Also in this connection, we find an explanation for the extraordinary and mysterious reflection attributed to Confucius:

Do you think that there must be the movements of the performers in taking up their positions, the brandishing of the plumes and fifes, the sounding of the bells and drums before we can speak of music? *To speak and carry into execution what you have spoken is ceremony; to act without effort or violence is music.*[34]

Everything is said in this mysterious formulation. Everything becomes clear if music is seen as an attribute of the sacrificial ceremony, which idealizes society and is tied to the monopolization of death, to the control over communication with the beyond, and, therefore, to prayer.

Without a doubt, music is a strategy running parallel to religion. The channeling power of music, like that of religion, is quite real and quite operative. Like an individual, a society cannot recover from a psychosis without reliving the various phases of its terror; and music, deep down, induces a reliving of noise's fundamental endowment with form, the channelization of the essential violence.

The musician: the sacrificed sacrificer; the worshipped and excluded Pharmakon; Oedipus and Dionysus. His work, which is political because it is religious, serves to integrate and channel anxiety, violence, and the imaginary, and to repress marginality. But in addition, because it is a threat of death, it transgresses; it heralds; it is prophetic of a new form of relations with knowledge and of new powers.

I believe that this hypothesis is new, even if some musicologists have taken steps in its direction. Adorno, for example, speaks superbly of music as a "*promise of reconciliation*," implicitly attributing to it the essential function of ritual sacrifice in all religious processes: reconciling people with the social order. Primordially, the production of music has as its function the creation, legitimation, and maintenance of order. Its primary function is not to be sought in aesthetics, which is a modern invention, but in the effectiveness of its participation in social regulation. Music—pleasure in the spectacle of murder, organizer of the simulacrum masked beneath festival and transgression— creates order. Every human production is in some way an intermediary and differential between people, and thus, in a sense, can be a channeler of violence.

I do not mean to say that music, at its origin, was only ritual in nature. Everything points to the fact that from the earliest times it has been present in every aspect of labor and daily life. It constitutes the collective memory and organizes society, often in ways less idealized than those described by Radcliffe-Brown:

> In the Andaman Islands everyone composes songs, and children begin to practice themselves in the art of composition when they are still young. A man composes his as he cuts a canoe or a bow or as he paddles a canoe, singing it over softly to himself, until he is satisfied with it. He then awaits an opportunity to sing it in public, and for this he has to wait for a dance. . . . He sings his song, and if it is successful he repeats it several times, and thereafter it becomes a part of his repertory. . . . If the song is not successful . . . the composer abandons it and does not repeat it.[35]

Music is lived in the labor of all. It is a collective selection process operating in the festival, the collective accumulation and stockpiling of code. But this does not contradict the essential function I would like to emphasize here: religion and the life of the collectivity form a whole, and the sacrificial molding of music gives meaning to its permanent presence in society. We are not talking about a secondary function. *Its fundamental functionality is to be pure order*. Primordially, and not incidentally, music always serves to affirm that society is possible.

The sacrificial codification of music, organized around a rigorous symbolism, eventually collapses. Internal and external noises do violence to the code and to the network. Simultaneously, money enters the picture and widens the rupture music contains within itself. The dynamics of codes is here premonitory: it implies and heralds a general mutation of the social codes, which the logic of the relations of music and money later reinforce, and which gradually extend from the empire of noise to that of matter.

A dynamic of codes, foreshadowing crises in political economy, is at work within music.

The Dynamic of Codes

Music brings into play tools (the voice, instruments), stockpiling points (griots [North or West African singer-poets], jongleurs, scores, records), and distribution networks (that is, a set of channels connecting the musical source to the listener). Thus under the sacrificial code, music was heard not only at the place of sacrifice, but also at all of the places of daily life and labor, where it was present in its ambiguous role as integrator and subverter. After the economy broke this network down, there came to be new modes of musical distribution, other places for music to be expressed, heard and exchanged, other networks. These networks changed music. But once again, the essential point is that they heralded a very thoroughgoing change in social organization, a change in the overall mode of economic production.

We can distinguish, conceptually, four essential types of networks and four possible forms of musical distribution, corresponding to the four fundamental structures it is possible to schematize. Each of these networks relate to a technology and a different level of social structuring. They succeeded and also interpenetrated one another in the economy of music. We will try to understand their structures, and the role money and the process of capitalist competition have played in their dynamic.

The Four Networks

The first network is that of *sacrificial ritual*, already described. It is the distributive network for all of the orders, myths, and religious, social, or economic relations of symbolic societies. It is centralized on the level of ideology and decentralized on the economic level.

A new network of music emerges with *representation*. Music becomes a spectacle attended at specific places: concert halls, the closed space of the simulacrum of ritual—a confinement made necessary by the collection of entrance fees.

In this network, the value of music is its use-value as spectacle. This new value simulates and replaces the sacrificial value of music in the preceding network. Performers and actors are producers of a special kind who are paid in money by the spectators. We will see that this network characterizes the entire economy of competitive capitalism, the primitive mode of capitalism.

The third network, that of *repetition*, appears at the end of the nineteenth century with the advent of recording. This technology, conceived as a way of storing representation, created in fifty years' time, with the phonograph record, a new organizational network for the economy of music. In this network, each spectator has a solitary relation with a material object; the consumption of music is individualized, a simulacrum of ritual sacrifice, a blind spectacle. The network is no longer a form of sociality, an opportunity for spectators to meet and communicate, but rather a tool making the individualized stockpiling of music possible on a huge scale. Here again, the new network first appears in music as the herald of a new stage in the organization of capitalism, that of the repetitive mass production of all social relations.

Finally, we can envision one last network, beyond exchange, in which music could be lived as *composition*, in other words, in which it would be performed for the musician's own enjoyment, as self-communication, with no other goal than his own pleasure, as something fundamentally outside all communication, as self-transcendence, a solitary, egotistical, noncommercial act. In this network, what is heard by others would be a by-product of what the composer or interpreter wrote or performed for the sake of hearing it, just as a book is never more than a by-product of what the writer wrote for the sake of writing it. At the extreme, music would no longer even be made to be heard, but to be seen, in order to prevent the composition from being limited by the interpretation— like Beethoven, brimming with every possible interpretation, reading the music he wrote but could no longer hear. Thus composition proposes a radical social model, one in which the body is treated as capable not only of production and consumption, and even of entering into relations with others, but also of autonomous pleasure. This network differs from all those preceding it; this capacity for personal transcendence is excluded from the other musical networks. In a society of ritualized sacrifice, representational speech, or hierachical and repetitive communication, egotistical pleasure is repressed and music has value only when it is synonymous with sociality, performance for an audience, or finally the stockpiling of "beauty" for solvent consumers. But when these modes of communication collapse, all that is left for the musician is self-communication. Here again, this network is ahead of its time and precedes a general evolution of social organization as a whole.

Order, Code, Network

In each network, as in each message, music is capable of creating order. Speaking generally and theoretically, in the framework of information theory, the information received while listening to a note of music reduces the listener's uncertainty about the state of the world. Euler even derived from this a definition of beauty as the faculty of discerning an order in a form: beauty as negentropy. In other words, the order created differs according to the network.

In the sacrificial network, music is part of a ceremony; it is a minor form of sacrifice, itself order-producing by nature.

In the network of representation, it is in general a flow of information, but it can create the conditions of a new order for the listener.

The repetition of music always creates disorder since it does nothing but replicate a recorded representation, imperfectly and without creating anything new: it is thus necessary in repetition to spend increasing amounts of value to maintain order.

Only the network of composition is order-producing by nature, for in the absence of any a priori code music creates an order for the observer.

Processes of Rupture: Order by Noise
and the Catastrophe Point

A network can be destroyed by noises that attack and transform it, if the codes in place are unable to normalize and repress them. Although the new order is not contained in the structure of the old, it is nonetheless not a product of chance. It is created by the substitution of new differences for the old differences. Noise is the source of these mutations in the structuring codes. For despite the death it contains, noise carries order within itself; it carries new information.[36] This may seem strange. But noise does in fact create a meaning: first, because the interruption of a message signifies the interdiction of the transmitted meaning, signifies censorship and rarity; and second, because the very absence of meaning in pure noise or in the meaningless repetition of a message, by unchanneling auditory sensations, frees the listener's imagination. The absence of meaning is in this case the presence of all meanings, absolute ambiguity, a construction outside meaning. The presence of noise makes sense, makes meaning. It makes possible the creation of a new order on another level of organization, of a new code in another network.

The idea that noise, or even music, can destroy a social order and replace it with another is not new. It is present in Plato:

This is the kind of lawlessness that easily insinuates itself unobserved [through music] . . . because it is supposed to be only a form of play and to work no harm. Nor does it work any, except that by gradual infiltration it softly overflows upon the characters and pursuits of men and from these issues forth grown greater to attack their business dealings, and from these relations it proceeds against the laws and the constitution with wanton license till it finally overthrows all things public and private. . . . For the modes of music are never disturbed without unsettling the most fundamental political and social conventions. . . . It is here, then, that our guardians must build their guardhouse and post of watch.[37]

This relates to the idea of rupture/rearrangement in the space of value: "In history as in nature, decomposition is the laboratory of life."[38]

But this *order by noise* is not born without crisis. Noise only produces order if it can concentrate a new sacrificial crisis at a singular point, in a *catastrophe*, in order to transcend the old violence and recreate a system of differences on another level of organization.

For the code to undergo a mutation, then, and for the dominant network to change, a certain catastrophe must occur, just as the blockage of the essential violence by the ritual necessitates a sacrificial crisis.

In other words, catastrophe is inscribed in order, just as crisis is inscribed in development. There is no order that does not contain disorder within itself, and undoubtedly there is no disorder incapable of creating order. This covers the dynamic of codes. There remains the question of the succession of noises and orders, and their interferences.

The Dynamic of Liquidation of Codes

Subversion in musical production opposes a new syntax to the existing syntax, from the point of view of which it is noise. Transitions of this kind have been occurring in music since antiquity and have led to the creation of new codes within changing networks. Thus the transition from the Greek and medieval scales to the tempered and modern scales can be interpreted as aggression against the dominant code by noise destined to become a new dominant code. Actually, this process of aggression can only succeed if the existing code has already become weak through use.

For example, the immense field of possible combinations opened by the tonal system led, in the course of two centuries, to excesses and the undermining of every tonal edifice. The gigantism of Romantic music announces and makes possible the end of this code as a tool for musical ordering. Later, the serial code, the last formalization of nineteenth-century determinism, was explored and then collapsed, liberating aleatory music. After that, there is no longer any fixed scale or dominant code. Every work creates its own linguistic foundation, without reference to fundamental rules.

Thus noise seems to reconstruct musical orders that become less and less coercive: the internal liquidation of codes began when music became an autonomous, constructed activity and was no longer an element of ritual. Parallel to mathematics, upon which musical codes are always very dependent, music evolved toward a pure syntax. For example, giant musical compositions were theorized when Fourier, then Helmholtz, decomposed sounds into infinite series of pure harmonic tones. The complete sound space, the image of sound particles, became theorizable when Markov constructed his theory of stochastic processes. But in spite of that, music even today contains innumerable forms that cannot be expressed in mathematics: dissonance, the mixture of order and disorder, syncopation, sudden change, harmonic redundancy. Each network pushes its organization to the extreme, to the point where it creates the *internal* conditions for its own rupture, its own noises. What is noise to the old order is harmony to the new: Monteverdi and Bach created noise for the polyphonic order. Webern for the tonal order. Lamont Young for the serial order.

But a noise that is *external* to the existing code can also cause its mutation. For example, even when a new technology is an external noise conceived as a reinforcement for a code, a mutation in its distribution often profoundly transforms the code. Although recording, for example, was intended first and foremost as a reinforcement and amplification of a preexisting speech mode, it in fact had an impact on the status of the contents of that speech; the network modifies the code within which messages are expressed. To take another example, the principle of printed reproduction impaired the authority of the preceding speech mode: inaugural, authentic, singular, manuscript-written speech. It even broke down the universal language of the time; conceived as a way of generalizing the use of Latin, printing instead destroyed it. The new medium and its network, which themselves became the universal language, the new form of the diffusion of authority, authorized a fragmentation of contents. In the sixteenth century, printing allowed the vernacular languages to spread and be reborn, and with them came a reawakening of the territorial nationalities. These deviations from the original usage of the code constitute a profound danger to the existing powers, so much so that they sometimes transform their morphologies in order to benefit from the new network themselves.

In music, the instrument often predates the expression it authorizes, which explains why a new invention has the nature of noise; a "realized theory" (Lyotard), it contributes, through the possibilities it offers, to the birth of a new music, a renewed syntax. It makes possible a new system of combination, creating an open field for a whole new exploration of the possible expressions of musical usage. Thus Beethoven's Sonata no. 106, the first piece written for the piano, would have been unthinkable on any other instrument. Likewise, the work of Jimi Hendrix is meaningless without the electric guitar, the use of which he perfected. But a new instrument does not always have the impact of opening a field of combination. The glass harmonica, invented by Benjamin Franklin and used by Mozart, and the arpeggione, used once by Schubert, have disappeared.

This double process of the rupture of codes (by internal and external noise) has destroyed, network by network, the socializing function of music. Music has not remained an "archipelago of the human" in the ocean of artifice that commercial society has become. The sound object itself has become artifice, independent of the listener and composer, represented, then repeated. Music used to cadence birth, labor, life, and death; it used to organize the social order. Today, it is too often nothing more than the consumption of past culture or a structure of universal mathematical invariants, a reflection of the general crisis of meaning. Communication has disappeared. We have gone from the rich priest's clothing of the musician in ritual to the somber uniform of the orchestra musician and the tawdry costume of the star, from the ever-recomposed work to the rapidly obsolescent object.

The ritual status of music has been modified by the network it subtends. It has become a simulacrum of the solitary spectacle of the sacrifice. The spectator has become an accomplice to individualized murder. Violence is no longer limited to the battlefield or the concert hall, but pervades all of society. The repetition of violence is no longer a promise of reconciliation. Once it has suppressed the mechanism of the collectively performed sacrifice, music does not "break the symmetry of reprisal . . . halt [this] recurrence by introducing something different. . . . Modern man has long lost the fear of reciprocal violence."[39]

After the process of identical repetition has extended to the whole of production, the end of differences unbridles violence and shatters all codes. Composition can then emerge. Composition, nourished on the death of codes.

Thus, there exist in the nature of codes mechanisms that entail their rupture and the emergence of new networks. But this destruction of codes is reinforced by the dynamic of the link between music and money. Each code or network entertains specific relations with money. Money creates, reinforces, and destroys certain of its structures.

Music and Money

When money first appeared, music was inscribed in usage; afterwards, the commodity entraps, produces, exchanges, circulates, and censors it. *Music is then no longer an affirmation of existence, it becomes valorized.* Its usage did not prevent its entering into exchange: since the time that societies' regulatory codes, prohibitions, and sacrificial rituals broke down, music has been unmoored, like a language whose speakers have forgotton the meaning of its words but not its syntax. However, no society can survive without order, in other words, the stable maintenance of differences, control over violence, and the channelization of unproductive expenditures. So either something has taken the place of music, despite its new status, or it has remained a tool of regulatory power, a fundamental form of the code of power functioning through new channels.

Music has become a commodity, a means of producing money. It is sold and consumed. It is analyzed: What market does it have? How much profit does it generate? What business strategy is best for it? The music industry, with all of its derivatives (publishing, entertainment, records, musical instruments, record players, etc.), is a major element in and precursor of the economy of leisure and the economy of signs.

At what stage in the production of a musical work is money produced? Does that stage vary according to the network and can the rupture of networks be explained by it? This economic problem is a complex one, because a musical work is an abstract form produced in several steps, which are structured differently depending on the the network within which the work is inscribed:

First, the composer produces a program, a mold, an abstract algorithm. The score he writes is an order described for an operator-interpreter.

Then, the interpreter creates an order in sound space with his instrument, which is a score-translating machine, a machine to decipher the coded thought of the composer. The form of the music is always influenced by the transmitter and the medium.

Finally, the object is produced, sold, consumed, destroyed, worn out. This involves a whole gamut of commodity production, from the score to the record. We will see that demand begins to require labor and has to be socially produced.

This production process treats commodities differently according to the network within which it takes place. The mode of production of exchange-value, the mode of accumulation of commodity-value, and the disequilibriums that lead to the rupture of the network are different for each one.

Every economic theory situates the production of money differently according to the form of its conceptualization.

This question, however, is of capital importance: the site of the creation of money in music explains the mode of growth, the power relations between the various actors, and the crisis affecting the entire network.

According to classical economics, a musical production creates wealth if it increases the real wages of the person who profits from it and makes him more efficient. In this sense, a singer creates monetary wealth if he increases the efficiency of his listeners, or if his performances lead to improved record sales. How productive he is is thus independent of his economic status. Similarly, a composer is considered productive if his work is marketed, regardless of his mode of remuneration.

Marxist political economy, on the other hand, locates the production of money solely in the manufacture of a material object by wage earners; as a result, a strange status is assigned to the composer in the process of the creation of commercial wealth.

In Marxist political economy, the question "How does music create wealth?" becomes "What kind of musical labor produces surplus-value?" This amounts to trying to determine what "productive" labor in music is, in other words, what kind of labor leads to the creation of value and the accumulation of capital.

Briefly, labor that contributes to the accumulation of capital, which creates surplus-value, is said to be *productive*, and labor is *unproductive* if it is only of interest to the purchaser for the use-value of its product. In the capitalist mode of production, only a wage earner who creates capital can be a producer. Therefore, by definition, and in contradistinction to classical economics, neither an unsalaried worker (as in the case of composers who are paid in the form of royalties), nor even a salaried worker who only implicitly creates capital by way of the consumption his activity presupposes or the inducement to produce he provides, can be considered productive. They create wealth in the capitalist mode of production while remaining outside it. Marx himself denied the productive nature of these two kinds of labor in the following words:

> [One cannot] present the labor of the pianist as indirectly productive, either because it stimulates the material production of pianos, for example, or because it gives the worker who hears the piano recital more spirit and vitality. Only the labor of someone who creates capital is productive, so any other labor, however useful or harmful it may be, is not productive from the point of view of capitalization; it is therefore unproductive. The producer of tobacco is productive, even though the consumption of tobacco is unproductive.[40]

Even in the case of a musician, value is produced if and only if he is a wage earner. Therefore, an activity that induces consumption or production does not in itself produce wealth. Similarly, handicrafts workers are never productive, because, Marx writes,

> They confront me as sellers of commodities, not as sellers of labor, and this relation therefore has nothing to do with the exchange of capital for labor; therefore also it has nothing to do with the distinction between *productive and unproductive labor*, which depends entirely on whether the labor is exchanged for money as money or money as capital. They therefore belong neither to the category of *productive* nor of *unproductive laborers*, although they are producers of commodities. But this production does not fall under the capitalist mode of production.[41]

Only the economic aspect of the distinction is significant: if the labor-power on the market is exchanged for capital, the labor creates value; if it is exchanged for money (revenue), the labor does not create value. Thus a musician who is paid a wage by someone who employs him for his personal pleasure is not a productive worker. His labor is exchanged for a wage or paid in kind: it is a simple exchange of two use-values. But if, for example, he plays a concert as the employee of someone in the entertainment business, he produces capital and creates wealth. In both cases, the composer of the score is unproductive. This distinction is uncharacteristic of capitalism.

Money Produced Outside Music: The Molder

The reader will have gleaned from the preceding classification discussion that the basic distinction to be made in understanding the production of money by music has to do not with the legal status of the musicians, but with the production network: the network of music is by no means exhausted by a definition of music's mode of distribution. The network also determines its mode of production and, beyond that, the mode of production of society as a whole. Thus we will see that the *economic laws of representation bear no relation to the laws of repetition* that are currently taking hold.

Let us first see how the economy of music functions in each economic network and where the creation of money takes place.

Music's mode of insertion into economic activity is different in each network. And this difference plays a major role in the transition from one network to another.

In the sacrificial network, music does not create wealth. The distinction between the productive and the unproductive is not even relevant to the shaman, nor to the jongleur later on. The minstrel, for his part, was unproductive, as has been most of the work of musicians like Wagner and Boulez up to the present day.

The accumulation of wealth through music appears only with representation, which both creates value and is the mode of functioning of the decline of the religious. In this network, value is created and accumulated outside the control of the composing musician. Marx himself provides a good analysis of the site of the creation of value in this network:

A singer who sings like a bird is an unproductive worker. When she sells her song, she is a wage earner or merchant. But the same singer, employed by someone else to give concerts and bring in money, is a productive worker because she directly produces capital.[42]

Therefore, the productive workers who create money are the performers, and the people who produce the instruments and the scores. But when the composer receives royalties on a work of his that is sold and represented, he remains curiously estranged from the wealth associated with him, since as an independent craftsman he is outside the capitalist mode of production. Good sense, however, requires that we recognize that he indirectly participates in the production of wealth in at least two ways: first, when productive workers (workers in music publishing) manufacture, using this stockpile of information and capital (in other words, using their past labor, another's past labor that has been appropriated by the entrepreneur, and their present labor), a commercial object (the score) whose sale to a musician (professional or amateur) realizes surplus-value; second, when the wage-earning musician, having acquired the score, represents the work.[43]

Nevertheless, the labor of the composer is not in itself productive labor, labor that is productive of commercial wealth. He is thus outside capitalism, at the origin of its expansion, except when even he is a wage earner selling his labor to capitalists (as is sometimes the case with film musicians). Generally remunerated with a percentage of the surplus-value obtained from the sale of the commercial object (the score) and its use (the performance), he is reproduced in every copy of the score and in each performance, by virtue of the royalty laws. His remuneration is therefore a kind of *rent*. A strange situation: a category of workers has thus succeeded in preserving ownership of their labor, in avoiding the position of wage earner, in being remunerated as a rentier who dips into the surplus-value produced by wage earners who valorize their labor in the commodity cycle. As the creator of the program that all of the capitalist production plugs into, he belongs to a more general category of people, whom I shall call *molders*. Entertainment entrepreneurs are capitalists; workers in publishing and performers are productive workers. Composers are rentiers. This situation is not without significance. It is even essential in understanding both the uniqueness of music and its prophetic nature in economic imitations.

If we go by the preceding analysis, a person whose creation originates an extensive process of material production is remunerated as a rentier: his income is independent of the quantity of labor he provides. Instead, it depends on the quantity of demand for that labor. He produces the mold from which an industry is built.

The musician is not an isolated case. There exist in the economy a considerable number of program producers, *molders*. In representation (handicrafts or archaic capitalism) each object is unique, and the mold is only used once. By contrast, in the economy of repetition a mold is used a great many times. If the remuneration of the molder is proportional to the number of sales, and not to the duration of his labor, then he can collect a rent and reduce the capitalist's profit. That is why it is in the interests of the capitalist process to incorporate molders as wage earners. Almost all of them have been integrated into large-scale research concerns; they have lost ownership of their creations and receive no income from their use by others. In this connection, the study of music is essential: if the molders of industry can in the future win the same rights as musical composers, and if the monopoly on the ownership of innovations can be replaced by remuneration based on their use by others, then it will possible for the results of the economy of music to be generalized. These results, as we shall see, are quite clear: *the specific remuneration of the composer has largely blocked the control of music by capital*; it has protected creativity and even today allows the relations of power between musicians and financiers to be reversed.

This becomes the essential question: is music an exception, or is it the herald of the reappropriation by all creators of their valorized labor? The spoils is capital. The outcome is uncertain.

We can now refine the analysis, distinguishing more precisely between the activities of representation and those of repetition. This distinction is not the same as the traditional distinction between industry and service, and it is far more important. Stated very simply, representation in the system of commerce is that which arises from a singular act; repetition is that which is mass-produced. Thus, a concert is representation, but also a meal à la carte in a restaurant; a phonograph record or a can of food is repetition. Other examples of representation are a custom-made piece of furniture or a tailored dress; and of repetition, ready-made clothing and mass-produced furniture. One provides a use-value tied to the human quality of the production; the other allows for stockpiling, easy accessibility, and repetition. In representation, a work is generally heard only once—it is a unique moment; in repetition, potential hearings are stockpiled.

Each of these modes of organization correspond to a very different logic, which we will study in detail in subsequent chapters. They are preceded by the sacrificial, noncommercial economy and are succeeded by the economy of composition.

These four kinds of modes of production interpenetrate in time and space, but it still seems possible to discern a certain economic logic of succession: *In music, as in the rest of the economy, the logic of the succession of musical codes parallels the logic of the creation of value.*

Representation emerged with capitalism, in opposition to the feudal world. It directed all new surplus-value toward the entertainment entrepreneur and the music publisher—few musicians made fortunes, and the royal courts lost their power. However, it was practically impossible to increase productivity. As the economy as a whole developed, the profit rate necessarily fell in this sector, due to increases in the cost of reproducing the labor force, which is determined on an economy-wide basis, and in the remuneration of authors of music.[44]

Capitalism then began to lose interest in the economy of representation. In addition, pricing problems arose. With radio and television, representation became available for free; it became impossible to charge for it as such. It is now hardly a capitalist activity anymore, except to the extent that it is used in record promotion.

Productive labor was able to concentrate its efforts in records, a new source of accumulation and a much broader one, and to produce a significant amount of value. At the same time, the new technology prolonged the process of the transformation of the author's labor into use-value. For this new process of production destroyed the conditions of earlier usage and required that a growing portion of surplus-value be used in the process of production itself—to pay the author's rent and to give meaning to the object being sold, to make the consumer believe that there was use-value in it, to promote demand. *In other words, in repetition a significant portion of the surplus-value created in the production of supply must be spent to create demand; and repetition produces less and less use-value.*

The economy of repetition can survive only if it is discovered each time how to recreate a use-value for the objects produced. Otherwise, meaning degenerates, like a mold worn out from overuse. Labor, then, serves to produce demand, and the essence of production becomes reproduction and the remuneration of rents. This creation of meaning is carried out by the industrial apparatus itself, and by the State apparatus: a portion of the collective machinery fulfills the function of producing demand, and, more radically, of producing the consumer. Accordingly, the extramarket production of demand becomes one of the market's conditions of existence. The need to spend surplus-value on circulation in order to preserve use-value can be interpreted as an inevitable loss of order inherent to any process of replication using a mold. We will see that the increasing difficulty of producing demand makes the production of supply more and more difficult. Thus repetition, like representation before it, is shattered by a crisis of usage and a decline in the creation of value.

In this crush of networks, the capitalist apparatus functions—more than as a machine for the accumulation of capital—as a machine of destruction reaching down to its own foundations, a machine to reroute the use of profit toward unproductive functions, to spend surplus-value on the payment of rents, to search out a meaning for its products. We have here a logic of the mutation of value-producing networks that is compatible with the logic of the movement of networks toward abstraction and the liquidation of all codes.

Capitalism thus realizes Marx's predictions so completely as to reduce the scope of Marxist analysis; it destroys its concepts, crushing them in capitalism's internal dynamic.

For this analysis also applies outside the framework of music. In the most modern sectors of our societies, exchange has destroyed usage, and surplus-value is spent to remunerate the producers of molds and to create a semblance of use-value for the objects that are mass-produced.

The process of this mutation in the location of value production, which we will describe in more detail in the three following chapters, makes music a herald of things to come. For after developing in music, it spreads throughout the entire economy, provoking a crisis the seriousness of which depends on the status of the molder and the efficiency of value creation.

The Advent of Simultaneity

Can it then be said that the way music makes money determines the evolution of the aesthetic code? That music depends on the economic status of musicians?

Marx wisely set the problem aside. ("As regards art, it is well known that some of its peaks by no means correspond to the general development of society; nor do they therefore to the material substructure, the skeleton as it were of its organization.")[45] However, the convergence we noted above, between the evolution of networks and the evolution of modes of production, in which music plays a prophetic role, obliges us to make a more thorough analysis.

Adorno, a musician and pessimist aristocrat, took Marxism as far as it could go on this question. For him, the evolution of music is a reflection of the decline of the bourgeoisie, whose most characteristic medium it is.[46] The interdependence between code and value, between musical style and economic status, is organized, he says, according to three fundamental principles. First: the relations of production set the limits of music; for example, the economic organization of capitalism was able, until Schoenberg, to hold back dissonances, the expression of the suffering of the exploited. Second: the composition, performance, and technique of music are forces of production. Finally, the relations of production and the forces of production are interdependent.

He deduces from this that the productive forces have less and less impact on the relations of production, and, in particular, that music has become increasingly separated from society. "Radical" music unmasks false musical consciousness and can transform the infrastructure, the relations of production outside the "sphere of music." Thus Wagner and Schoenberg are autonomous negations, outside the dictated rules. "It is only in dissonance, which destroys the faith of those who believe in harmony, that the power of seduction of the rousing character of music survives."[47]

Today, with the *superimposed* presence of networks translating different codes of production, the unicity of both the musical order and the economic order is gradually disappearing. The aesthetic networks have broken down and the modes of production have changed. But they are no longer unique to a certain period. In the classical age, the tonal order of music seems to have imposed itself as a stable norm. The stylistic reversal that occurred with the emergence of the rigorous code of harmony, which is proper to the network of representation, was brutal and unnuanced. A few years were sufficient for the new style to take hold and for the musicians and styles of the past to be forgotten. Haydn's style imposed itself on all of his contemporaries, musicians like Stamitz, Richter, Holzbauer, Filitz, Toeschi, Danzi, Cannabich, Wagenseil, Rosetti, and Mozart. At the same time, a new economic organization of society was established with its own dominant code, that of representation, which would produce the star: only Haydn and Mozart would be remembered by history. Representation became dominant and replaced the sacrificial network in practice.

Afterward, in codes as in value, the domination of one of the networks no longer excludes the existence of the others, as flotsam from the past or fragments of the future. There is no longer any sudden rupture, but instead the successive dominance of certain codes and networks over others still present.

Just as it would be imprecise to hold that the competitive order made way for the monopolistic order, or that a socialist society is born the moment the means of production have been collectively appropriated, the reality is much more complex, and music makes this simultaneity audible.

> There is no universal capitalism, no capitalism in itself; capitalism is at the crossroads of all kinds of formations. It is always by nature neo-capitalism; it invents, for the world, its eastern face and its western face and its reshaping of both.[48]

In the same way that today's socialisms have remained pregnant with feudalism and capitalism, so also do representation, repetition, and composition give mutual sustenance.

The mechanism of their substitution is everywhere adrift. In the liquidation of codes and value, there is no direction or meaning, no project discernible to the observer. All the observer sees is noise in relation to the code, a crisis in relation to value. That is why a genealogy of music is impossible. All attempts at one are nothing but immense theoretical mystifications, and their success is due only to the general anxiety in the face of the meaningless. In the simultaneity of networks, there is nothing but a conflict of codes and institutions, a drift toward the absence of meaning.

Just as what is essential in a philosophy is not in what it says, but in what it does not say, the future of an organization is not in its existence, but in its opposite, which reveals its mutation. Today, the future is in our lacks, our suffering, and our troubles: repetition expresses the negative image of this absence of meaning, in which the crisis now in process will crystallize through a multiplication of simultaneous moments, the independent and exacerbated presence of the past, present, and future. To write history and fashion political economy is then to describe this cascade, to describe its instability and movement, more profound than the logic of each code, where a meaning beyond nonsense may be reborn, a sound beyond noise.

It is no accident that Carlos Castaneda noticed this profound relation between drugs, knowledge, and the flow of water:

> Go first to your old plant and watch carefully the watercourse made by the rain. By now the rain must have carried the seeds far away. Watch the crevices made by the runoff, and from them determine the direction of the flow. Then find the plant that is growing at the farthest point from your plant. All the devil's weed plants that are growing in between are yours. Later, as they seed, you can extend the size of your territory.[49]

This metaphor contains one of the greatest lessons in scientific methodology audible today: it is senseless to classify musicians by school, identify periods, discern stylistic breaks, or read music as a direct translation of the sufferings of a class. Music, like cartography, records the simultaneity of conflicting orders, from which a fluid structure arises, never resolved, never pure. How can one act upon a stream laden with so many colliding temporalities? How can one, at the same time, account for this foreshadowing of crisis in the degradation of the code and in the displacement of the production of money in music?

Today, however, a single code threatens to dominate music. When music swings over to the network of repetition, when use-time joins exchange-time in the great stockpiling of human activity, excluding man and his body, music ceases to be a catharsis; it no longer constructs differences. *It is trapped in identity and will dissolve into noise.*

Violence, then, threatens more than ever to sweep across a meaningless, repetitive, mimetic society. Outlined on the horizon is the real crisis, the great crisis, the *crisis of proliferation* accompanied by the absolute dissolution of the place of sacrifice, of the arena of political action, and of subversion. But at the same time, the loss of meaning becomes the absence of imposed meaning, in other words, meaning rediscovered in the act itself—composition: in which there is no longer any usage, any relation to others, except in the collective production and exchange of transcendence. But composition necessitates the destruction of all codes. Is it to be hoped, then, that repetition, that powerful machine for destroying usage, will complete the destruction of the simulacrum of sociality, of the artifice all around us, so that the wager of composition can be lived? Must Lent supplant Carnival for the conditions of the Round Dance to emerge?

Chapter Three
Representing

Make people believe. The entire history of tonal music, like that of classical political economy, amounts to an attempt to make people believe in a consensual representation of the world. In order to replace the lost ritualization of the channelization of violence with the spectacle of the absence of violence. In order to stamp upon the spectators the faith that there is a harmony in order. In order to etch in their minds the image of the ultimate social cohesion, achieved through commercial exchange and the progress of rational knowledge.

The history of music and the relations of the musician to money in Europe since the eighteenth century says much more about this strategy than political economy, and it says it earlier.

Beginning in the eighteenth century, ritualized belonging became representation. The musician, the social memory of a past imaginary, was at first common to the villages and the court, and was unspecialized; he then became a domiciled functionary of the lords, a producer and seller of signs who was free in appearance, but in fact almost always exploited and manipulated by his clients. This evolution of the economy of music is inseparable from the evolution of codes and the dominant musical aesthetic. Although the economic status of the musician does not in itself determine the type of production he is allowed to undertake, there is a specific type of musical distribution and musical code associated with each social organization. In traditional societies, music as such did not exist; it was an element in a whole, an element of sacrificial ritual, of the channelization of the imaginary, of legitimacy. When a class emerged whose power was based on commercial exchange and competition, this stabilized system of

46

musical financing dissolved; the clients multiplied and therefore the distribution sites changed. The servants of royal power, despite the occasional efforts of revolutionary institutions, were no longer in the service of a singular and central power. The musician no longer sold himself without reserve to a lord: he would sell his labor to a number of clients, who were rich enough to pay for the entertainment, but not rich enough to have it to themselves. Music became involved with money. The concert hall performance replaced the popular festival and the private concert at court.

The attitude toward music then changed profoundly: in ritual, it was one element in the totality of life; in the concerts of the nobility or popular festivals, it was still part of a mode of sociality. In contrast, in representation there was a gulf between the musicians and the audience; the most perfect silence reigned in the concerts of the bourgeoisie, who affirmed thereby their submission to the artificialized spectacle of harmony—master and slave, the rule governing the symbolic game of their domination. The trap closed: the silence greeting the musicians was what created music and gave it an autonomous existence, a reality. Instead of being a relation, it was no longer anything more than a monologue of specialists competing in front of consumers. The artist was born, at the same time as his work went on sale. A market was created when the German and English bourgeoisie took to listening to music and paying musicians; that lead to what was perhaps its greatest achievement—freeing the musician from the shackles of aristocratic control, opening the way for the birth of inspiration. That inspiration was to breathe new life into the human sciences, forming the foundation for every modern political institution.

Representation, Exchange, and Harmony

From the Musician-Valet to the Musician-Entrepreneur

The minstrel, a functionary, only played what his lord commanded him to play. As a valet, his body belonged entirely to a lord to whom he owed his labor. If his works were published, he would receive no royalty, nor was he remunerated in any way when others performed his works. A piece in the ideological apparatus, charged with speaking and signifying the glory of the prince—a simulacrum of the ritual—he would compose what the lord ordered him to compose, and the lord had use of and ownership over both musician and music.

The court musician was a manservant, a domestic, an unproductive worker like the cook or huntsman of the prince, reserved for his pleasure, lacking a market outside the court that employed him, even though he sometimes had a sizable audience. Bach's work contract, for example, is that of a domestic:

Whereas our Noble and most gracious Count and Master, Anthon Günther, one of the Four Counts of the Empire, has caused you,

Johann Sebastian Bach, to be accepted and appointed as organist in the New Church, now therefore you are, above all, to be true, faithful, and obedient to him, His above-mentioned Noble Grace, the Count, and especially to show yourself industrious and reliable in the office, vocation, and practice of art and science that are assigned to you; not to mix into other affairs and functions; to appear promptly on Sundays, feast days, and other days of public divine service in the said New Church at the organ entrusted to you; to play the latter as is fitting; to keep a watchful eye over it and take faithful care of it; to report in time if any part of it becomes weak and to give notice that the necessary repairs should be made; not to let anyone have access to it without the foreknowledge of the Superintendent; and in general to see that damage is avoided and everything is kept in good order and condition. As also in other respects, in your daily life to cultivate the fear of God, sobriety, and the love of peace; altogether to avoid bad company and any distraction from your calling and in general to conduct yourself in all things toward God, High Authority, and your superiors, as befits an honor-loving servant and organist. For this you shall receive the yearly salary of 50 florins; and for board and lodging 30 talers.[50]

The same features are found in the majority of musicians' work contracts of the period. Another example in Haydn's contract with Prince Esterházy, signed May 1, 1761, which makes him a conductor, composer, administrator, and the inheritor of his predecessor's debts.[51] The contract thus constitutes a relation of domesticity and not one of exchange.

Tool of the political, his music is its implicit glorification, just as the dedicatory epistle is its explicit glorification. His music is a reminder that, in the personal relation of the musician to power, there subsists a simulacrum of the sacrificial offering, of the gift to the sovereign, to God, of an order imposed on noise. Lully, in his dedication of *Persée* to Louis XIV, writes:

It is for Your Majesty that I undertook this work, I must dedicate it only to you, Sire, and it is you alone who must decide its destiny. The public sentiment, however flattering it may be for me, does not suffice to make me happy, and I never believe I have succeeded until I am assured that my work has had the good fortune of pleasing you. The subject seemed to me of such beauty that I had no difficulty developing a strong fondness for it, I could not fail to find in it powerful charms; you yourself, Sire, were kind enough to make the choice, and as soon as I laid eyes upon it, I discovered in it the image of Your Majesty. . . . I well know, Sire, that on this occasion I should not have dared publish your praise; not only for me is your praise too elevated a topic, it is beyond even the reach of the most sublime eloquence. However, I realize that in describing the true gifts Perseus received from the Gods, and the astonishing deeds he so gloriously accomplished, I am tracing a portrait of the heroic qualities and prodigious actions of Your Majesty. I feel that my zeal would run away with me if I neglected to restrain it.[52]

Never has the political discourse of music seemed so strong, so linked to feudal powers, as in this period, when Molière had his music master say "Without music no State could survive,"[53] when the cord attaching music to royal power was so strong.

However, cracks were starting to form and irony was beginning to show through the praise. The domestic, knowing that he could begin to depend on other economic forces than the courts, wanted to be done with this double language of order and subversion. The philosophers of the eighteenth century, moreover, provided a political ally and ideological foundation for the revolt of the artist against his guardian, for the will to artistic autonomy. Marmontel, in one of his furious articles in the *Encyclopédie*, violently attacks musicians who play the game of the dedicatory epistle, which was a symbol of submission to the feudal world, an occasion for the domestic to beg for his reward:

The signs of kindness one boasts of having given, the favorable welcome he made perceptible, the recognition that moves one so, and about which he is so surprised; the part that one is supposed to have had in a work that put him to sleep when he read it; his approval, the often imaginary history of which one recounts to him; his fine actions and sublime virtues, left unmentioned for good reason; his generosity, which one praises in advance, etc. All of these formulas are stale.[54]

They had grown stale because the economic role of the epistle had diminished. At the end of the eighteenth century, it was no longer praise sung for the lord and master, but one way among others for an independent artist to obtain funds from a financial, feudal, or capitalist power. The epistle was thus the last link between music and a declining feudal world whose domination of art would soon come to an end, at least in France. Artists even wrote to the mighty with an impertinence that antecedes and foreshadows the political rebellion of the bourgeoisie.

In 1768, Grétry wrote sarcastically to the count of Rohan Chabot: "I also request that you give me a flat refusal, if you have reason to do so, for example, to avoid the crowd of importunate authors who doubtless seek to dedicate their works to you."[55] This same Grétry later wrote, with very great analytic insight: "I saw the birth and realization of a revolution by artist musicians, *which came a little before the great political revolution.* Yes, I remember: musicians, maltreated by public opinion, suddenly rose up and repulsed the humiliation weighing down upon them."[56]

Of course, patronage did not disappear with the eighteenth century; it still exists, as we shall see. But music was already in contact with a new reality; it refused to stay tied to a camp whose power was dwindling. It ceased to be written solely for the pleasure of the idle and became an element in a new code of power, that of the solvent consumer, the bourgeoisie. It became an element of social status, a recollection of the hierarchical code whose formation it encoded. In the beginning, it was a mere possession: *one "had" music, one did not listen to it.* Mozart, writing of his reception in Paris in 1778, described this moment of transition in the status of music as "detestable," because music was no longer the sign of power now lost, and not yet the abstract sign of a new force: "So I had to play to the chairs, tables, and walls."[57]

Everything was reversed: art was no longer a support for feudal power; the nobility, after it lost the ability to finance music, still tried for a time to use its culture to legitimate its control over art:

> The count, afterwards the duke of Guines, played the flute exceedingly well. Vendelingue, the greatest flautist of the time, conceded that they were of equal skill. The count of Guines had a whim to play with him one evening in a public concert; they played twice, alternating on first and second flute, and with exactly the same success.[58]

Similarly, in 1785, Prince Lebkowitz, who had been appointed regent in 1784, played second violin in a quartet formed by his chapel master, Wranitzky, who inspired Beethoven's quartets. This prince was also one of three patrons who, after 1809, guaranteed Beethoven an annual allowance. Still others tried to go into business, for example, the count of Choiseul, who conceived the project of making the Opéra-Comique profitable by means of a commercial gallery. Thus before the transition in political institutions from divine right to political representation, the rupture had already taken place in music.

The first concerts to draw a profit took place in London in 1672; they were given by Bannister, the violinist and composer. Entrepreneurs organized concerts for the bourgeoisie, in whose dreams they were a sign of legitimacy. The concert hall appears at this time as the new site of the enactment of power. Up until the seventeenth century, music financed by princes was heard in churches and palaces; but now it was necessary to make people pay to hear music, to charge for admission. The first concert hall we have record of was established in Germany in 1770 by a group of Leipzig merchants who began in an inn called "Zu den drein Schwanen" ("At the Three Swans"); then in 1781 they converted a clothier's shop into a concert hall—music was literally confined within the walls of commerce.

Handel, one of the first composers to seek financial support outside the royal courts, described this transition as it was taking place in England. In a letter written in 1741, considerations of decorum prevail over his concern for profit:

> The Nobility did me the Honour to make amongst themselves a Subscription for 6 Nights, which did fill a Room of 600 Persons, so that I

needed not sell one single Ticket at the Door. . . . . The Audience being composed (besides the Flower of Ladyes of Distinction and other People of the greatest Quality) of so many Bishops, Deans, Heads of the Colledge, the most eminents [*sic*] People in the Law as the Chancellor, Auditor General, &tc. all which are very much taken with the Poetry. So that I am desired to perform it again next time. I cannot sufficiently express the kind treatment I receive here. . . . . They propose already to have some more Performances, when the 6 Nights of subscription are over, and My Lord Duc the Lord Lieutenant (who is allways present with all His Family on those Nights) will easily obtain a longer Permission for me by His Majesty.[59]

In France, the opposition of Lully and the Royal Academy of Music, which monopolized all power in the name of the king and prohibited the diffusion of music, was not overcome until 1725. Philidor's *Concert spirituel* ("Spiritual Concert") followed by Gossec's *Concert des Amateurs* ("Concert of Amateurs") in 1769, brought the music of power to the bourgeoisie of France, after that of England. The privilege granted the Royal Academy of Music was in fact so exclusive that no one could organize a performance for which admission would be charged or a public ball without the authorization of the director. This was taken to such an extreme that Italian comedians were ordered to pay a fine of 10,000 livres to the Royal Academy of Music for including song and dance in a public performance of *Fêtes de Thalie* ("Festivals of Thalia"); the following year, they were fined 30,000 livres for adding ballet to *La Fête Ininterrompue* (The Continuous Festival"), *Nouveau Monde* ("New World"), and *L'Inconnue* ("The Unknown Woman").

Everything changed once this monopoly was broken. The musician received a new status, causing a shake-up in the economic status of the musical work and the entire economy of music. When music entered the game of competition, it became an object from which income could be drawn without a monopoly; it fell subject to the rules and contradictions of the capitalist economy.

The Emergence of Commodity Music

In order for music to become institutionalized as a commodity, for it to acquire an autonomous status and monetary value, the labor of the creation and interpretation of music had to be assigned a value. Next—and this happened much later—it was necessary to establish a distinction between the value of the work and the value of its representation, the value of the program and that of its usage.

This valorization of music took place in opposition to the entire feudal system, in which the work, the absolute property of the lord, had no autonomous existence. It was constructed on the basis of the concrete existence, in an object (the score) and its usage (the representation), of a possible commercial valorization. Music, then, did not emerge as a commodity until merchants, acting in the name of musicians, gained the power to control its production and sell its usage, and until a sufficiently large pool of customers for music developed outside the courts, for which it had been formerly reserved. The history of copyright in France, where creators' ownership over signs was first affirmed, is fascinating in this context. In the beginning, the purpose of copyright was not to defend artists' rights, but rather to serve as a tool of capitalism in its fight against feudalism. Before a September 15, 1786, ruling of the Conseil du Roi ("The Council of the King"), musical composers had no control over the sale or representation of their works, with the exception of operas; in principle, only the director of the Royal Academy had such rights. In fact, the music publishing industry grew out of the bookmaking industry, itself the result of the existence of a market for books. Even the law protecting books had not been easily elaborated, since it ran counter to the interests of the copyists, who had controlled the production of copy-representations of writing until the discovery of printing. The copyists, who had a monopoly over the reproduction of all manuscripts regardless of type (text or score), were for a time successful in opposing printing, which created the foundation for repetition and the death of representation in writing. They obtained, by decision of the Parlement of Paris, authorization to destroy the presses. Their success was fleeting. Louis XI, who had need of a press to assure a wide audience for Arras' treatise, annulled the decision, and bookmakers were granted privileges for literary publishing for the first time.

The political repercussions were immense: the printing press was the downfall of the fixed word of power, proposing a schema of generalized reproduction in its stead. It destroyed the weight of the original. It detached the copy from its model. A distribution technique that began as a harmless support for a certain system of power ended up shattering it instead.

In music, printing gave meaning to the advent of polyphony and the scale, in other words, the advent of harmonic writing and standardized scores. The publisher created a commercial object, the score, to be sold by the lord, not the musician. Then in 1527, music publishing received the same rights accorded literary publishing; that is, the publisher of the work was given exclusive rights over its reproduction and sale. But this privilege was still limited to the material reproduction of the score and did not apply to the work itself, which was unprotected against piracy. In particular, all popular music was excluded from copyright protection and would remain so until the nineteenth century: in the absence of a solvent market, it occurred to no one to commodify it or protect its ownership. Thus a musician or his master could sell a work, whether a song or an instrumental work, just like any other possession. But once it was sold, it belonged to the publisher, who could market it as he saw fit, with no possibility of the musician's opposing it. Copyright thus established a monopoly over reproduction, not protection for the composition or control over representations of it. In the beginning, the author only had control over written reproductions; this is an indication of how little weight was given to performances. The space of communication was strictly limited to printed characters. Outside the written reproduction, valueless.

Of course, certain especially well-known musicians obtained privileges for certain of their works at a very early date. But they had little impact, and the monopoly of the copyists was replaced by a monopoly of the publishers. The publishers, organized as a guild, had the exclusive right to print and sell scores and, if they wished, to combat piracy, imitations, plagiarism, and unauthorized performances. Writers and composers were thus totally dependent upon their publishers. The creator, whether a salaried worker or an independent, minstrel or jongleur, remained powerless in the face of the transformation of his labor into a commodity and money.

In France, two publishers dominated the market for a century: Le Roy and Ballard. They became partners in 1551, and on February 16, 1552, were granted both the permanent privilege to publish any vocal or instrumental music for which earlier privileges had expired, and the privilege to be the exclusive publishers of the king's music. Unlike booksellers, who could live off their backlists, music publishers did not have numerous scores of earlier music to exploit, since the language of music had just stabilized. Therefore, they were quick to publish, for the benefit of provincial notables and with the court's consent, the works of contemporary musicians, who thus fell totally under their control.

But publishers' control, which barred the musicians from receiving any compensation, did not outlast the status of the minstrel. Little by little, as they dissociated themselves from the courts, musicians obtained part ownership of their labor; in other words, they succeeded in separating ownership of the work from the object manufactured by the publisher—even though they sold the right to publish it, they retained ownership of it and control over its usage. In addition, publishers in the provinces revolted against the monopoly granted to Paris to publish and sell a work over the whole of France. Although music publishing was not at the forefront of the struggle against the Parisian monopoly, it reaped the benefits. Thus emerged—and this is of capital importance—the immateriality of the commodity, the exchange of pure signs. Even though the written form was to remain for a long time the only form, the only reality of music that could be stockpiled, the sign was already for sale.

Little by little, the power of the publishers was dismantled. First, the Conseil du Roi ruled on August 13, 1703, that any privilege that had been granted for an indefinite period was legally void. Then in 1744, the provincial publishers won, in the form of an extension to the provinces of the bookseller's statute of 1723, an end to the monopoly Parisian publishers enjoyed over the publication of a work. In battling their competitors at the center, the reproducers of the periphery pioneered the concept of the work, thus serving the interests of the composers.

Finally, authors and composers succeeded in winning from the publishers a portion of the revenue drawn from the sale of publications and performances to the bourgeoisie.

Lully, despite his omnipotence, was unable to obtain the right to publish his own works from the royal powers, who were at the time very responsive to the interests of the publishers. A ruling of June 11, 1708, explicitly states that he does not have the right to publish his own works or draw income from them. The situation turned around a bit in 1749, when Louis XV refused to issue the Ballard press a general privilege for music engraving. This constituted a mutation in the balance of power: the musician earned a new share in the ownership of the work. The work as commodity became separated from its material support. Control over its sale and unauthorized use was explicitly granted to the musicians themselves. At least, this was the case for what the law chose to designate as a "work" of music, which at the time meant a score of sufficient stature to be performed before a solvent audience, not songs or works destined for a popular audience, in other words, an audience not confined in a concert hall.

On March 21, 1749, the Conseil du Roi recognized the intangibility of musical works, and a ruling in 1786 finally formulated a general regulation, still in effect: since "the piracy of which the composers and merchants of music were complaining was injurious to the rights of artists and to the progress of the arts, and ownership rights were daily becoming less respected, and the talented were deprived of their productions," it was decided that the privilege of the seal required for publication under the terms of the bookselling laws would "only be granted to commercial publishers after they have justified the transfer of rights that will be made to them by the authors or owners." The regulation also specified the form and terms of the declarations and registration necessary to assure ownership rights, and lastly prohibited "under penalty of a fine of 3,000 livres, the unauthorized use of any piece of music, as well as of engravers' stamps and trademarks." Thus the ownership rights of authors over their "works" was finally recognized.

Shortly afterward, the Revolution began; it at first merely codified this protection granted to the property of composers, as independent entrepreneurs, against the capitalist publishers. During a particularly troubled period, it in effect reenacted all of the measures promulgated by the courts of the monarchy: a law of January 13-19, 1791 and a decree of July 19-24, 1793 prohibited the pirating of musical "works" and performances not authorized by the author. This test regulated the musical economy of representation. It assigned the publishers the function of valorizing music, the ownership of which remained with the composers.

But out of a desire to extend the protection given to music and musicians against the effects of money, the French Revolution later tried to nationalize music; it was an incredible attempt (possibly unique in history) at rationally organizing the production of music, a project conceived as a voluntary apparatus for the elaboration of a State ideology and for the standardization of cultural production.

The Centralized Planning of Music

In the same year, 1793, the Convention tried to transfer ownership of music to the State, and to return to a political control of music even more extreme than that practiced under royal power. In this, music is indicative of the contradictions of the period, of the instability of the control exerted over social relations, and of the simultaneity of incoherent political projects at the end of the eighteenth century. On the one hand, the Revolution sanctioned the conquests the bourgeoisie made in the eighteenth century and affirmed the right of the individual to ownership of his labor; on the other hand, it returned control over ideological production to national and popular State political power, with the explicit goal of resisting the bourgeoisie and the dangers money poses for music.

The history of this attempt to construct a centrally planned and monopolistic tool for the production of music is edifying, and undoubtedly unique in the worldwide history of music.

In 1793, the National Convention created the National Institute of Music in a move to give the revolutionary State rights over musical production still more totalitarian than those the royal State ever had at its disposal.

The objective of this institution, according to its impassioned proponent Gossec, was to assemble the "premier artists of Europe"[60] in the category of wind instruments—about three to four hundred musicians—and put them to work "annihilating the shameful torpor into which [the arts] have been plunged by the impotent and sacrilegious battle of despotism against liberty."[61] The aim was to produce music that would "support and bestir, by its accents, the energy of the defenders of equality, and to prohibit that music which softens the soul of the French with effeminate sounds, in salons or temples given over to imposture."[62] The role of music, he said, is to glorify the Republic at celebrations, to take music where the people can hear it, rather than immuring it at the places of power: "Our public squares will henceforth be our concert halls."[63] Leave the church and palace, those feudal inventions, behind—but also the concert hall, the invention of the bourgeoisie.

This musical corps was to create works of music and perform them at public celebrations throughout the Republic. To make its revolutionary vocation all the more clear, it was incorporated into the National Guard, created in 1789. Music was thus intended to be the guardian of the State against the bourgeoisie itself: "Then the nation will more easily create the kind of musical corps which rouse our republican phalanxes to battle."[64]

The Institute was created in Messidor, year II, along very bureaucratic lines. It was placed under the direction of some of the best musicians of the time

(Mehul, Cherubini, Gossec), and was divided into schools, of 80 students each, with a total of 115 instructors. The only things banned were the chant and the harpsichord, which were undoubtedly too closely associated with the ancien régime. The Institute, then, was called upon to "regulate all music everywhere, to arouse the courage of the defenders of the fatherland and increase the ability of the *départements* to add pomp and appeal to civil ceremonies."[65] Each month, it had to furnish the Public Health Committee at least one symphony, one hymn or chorale, one military march, a rondeau or quick march and at least one patriotic song, all of which added up to a 50-60 page notebook, 550 copies of which had to be made; it also had to send 12,000 copies of patriotic hymns and songs to the various armies of the Republic each month. The Institute's musicians were salaried and had no rights over performance of the works. Finally, the Institute was given two economic objectives. First, "it will effect the naturalization of the wind instruments, which we are obliged to import from Germany; this neutralizes an important branch of industry in France, deprives a portion of the large population of the Republic of its livelihood,"[66] and causes musicians to leave the country. Second—and this is the essential idea behind the institution—it was a way for the revolutionary State to prevent the bourgeoisie from laying hold of music and debasing it: *"Who then will encourage the useful sciences, if not the government, which owes them an existence that in times past was procured for them by the rich and powerful, amateurs in taste and tone? Can we overlook the fact that the new rich, who emerged from the dregs of the Revolution, are crapulous and ignorant, and only propagate the evils produced by their insatiable and stupid cupidity?"*[67]

This revolutionary dream of preventing money from dictating the course of music soon crumbled. Despite the fact that the Institute was retained (after 1795, under the name of the Conservatory), music was to become the property of the "new rich," and what could be characterized as a project in *militaro-musical nationalism* was to collapse as industry expanded. The bourgeoisie would organize the essentials of production and representation, and control musical inspiration for the length of the nineteenth century.

The Conservatory, in the beginning generously funded, quickly became a low priority. As early as year V, its administrators had to resist a proposal to reduce its support; in year VIII, after a violent campaign was waged against its productions, they were unable to prevent the reductions. Under the Empire, the Conservatory increased in size again. But in the Restoration period, the title Royal School, originally given to the Academy of Music in 1784, was readopted; its sole function was to supply music for the Opera. All other music was thus abandoned to commercial exchange. The bourgeoisie and publishers would control its commercialization, limiting protection to works consumed by the bourgeoisie. For example, the Penal Code of 1810 only protected "dramatic works," and thus excluded from protection popular songs and music, unless a judge ruled otherwise.[68] Bourgeois law for bourgeois music—music that was rarely played, and the piracy and performance of which was easily controllable.

In representation, the musician no longer sold his body. He ceased to be a domestic, becoming an entrepreneur of a particular kind who received a remuneration from the sale of his labor. The musician's economic status and political relation with power changed in the course of the great political upheaval of the time, as did the aesthetic codes and forms in which the new audience wished to see itself reflected.

Thus delimited, music became the locus of the theatrical representation of a world order, an affirmation of the possibility of harmony in exchange. It was a model of society, both in the sense of a copy trying to represent the original, and a utopian representation of perfection.

The channelization of violence became more subtle, since people had to content themselves with its spectacle. It was no longer necessary to carry out ritual murder to dominate. The enactment of order in noise was enough.

Music as the Herald of the Value of Things

Thus a change in the nature of listening changed the code: up until that time, music was written not to be represented, but to be inscribed within the reality of a system of power, to be heard as background noise in the daily life of men. When people started paying to hear music, when the musician was enrolled in the division of labor, it was bourgeois individualism that was being enacted: it appeared in music even before it began to regulate political economy. Until the eighteenth century, music was of the order of the "active"; it then entered the order of the "exchanged." Music demonstrates that exchange is inseparable from the spectacle and theatrical enactment, from the process of *making people believe*: the utility of music is not to create order, but to make people believe in its existence and universal value, in its impossibility outside of exchange.

Music makes audible an obvious truth which, though never explicit and too long forgotten, has formed the foundation for all political thought since the eighteenth century: *the concept of representation logically implies that of exchange and harmony. The theory of political economy of the nineteenth century was present in its entirety in the concert hall of the eighteenth century, and foreshadowed the politics of the twentieth.*

As we have seen, charging admission for representation presupposes the sale of a service, in other words, the expression of an equivalence between musical production and other commercial—and no longer domestic—activities. This idea of the exchangeability of music is disruptive, because it places music in the context of abstract, generalized exchange, and consequently of money.

This new context is of considerable theoretical consequence.

On the one hand, representation entails the idea of a model, an abstraction, one element representing all the others. It thus relates to the spectacle of the political and the imaginary, but first of all to money, the abstract representation of real wealth and the necessary condition for exchange. The idea, which was very new at the time, that it is possible to represent a reality by a form, a semantics by a syntax, opened the way for scientific abstraction, for the attainment of knowledge through mathematical models.

On the other hand, the entrance of music into exchange implicitly presupposes the existence of an intrinsic value in things, external and prior to their exchange. For representation to have a meaning, then, what is represented must be experienced as having an exchangeable and autonomous value, external to the representation and intrinsic to the work.[69] Western music, in creating an aesthetic and instituting representation, implies the idea of a value in things independent of their exchange, prior to their representation. The work exists before being represented, has a value in itself. This brings up the central problem of political economy, that of measuring the value of things, and gives us an insight into the Marxist response: for labor is the only common standard for all of these representations, and it is the labor of the musician that forms the basis of their value.

Representation requires a *closed framework*, the necessary site for this creation of wealth, for the exchange between spectators and productive workers, for the collection of a fee. Music, meaningless outside of religion, takes root in representation, and therefore in an exchange of labor allowing a comparison between representations to be made. Music is judged with reference to the musical code that determines its complexity. This gives us insight into the entire labor theory of value: the schema for the determination of ticket prices for a representation requires the comparability of musical works according to criteria external to their representation, in other words, the existence of a standard for determining an autonomous value of the spectacle. This standard can only be the labor of the musicians. Thus by replacing barter with a standard, representation empties exchange-time and, in an extreme irony, the very person who is remunerated as a rentier is the one who provides the insight into labor-value as the standard for capitalist exchange.

The enactment of music by the bourgeoisie, represented and then exchanged according to criteria of deritualized usage, thus contains in embryonic form the entirety of nineteenth-century political economy, particularly that of Marx—to be exact, the theory of exchange, and the most solid foundation for value as the labor incorporated in the object. It also implies the existence of ineluctable laws, like a score unfolding before each man and each class, spectators of the contradictions of society. Music announces—shouts it out, even—that the political economy of the nineteenth century could only be theater, a spectacle trapped by history.

But at the same time as music appears as having a value outside exchange, at the same time as it announces exchange as the transformation of value into money, it designates this standard as indefensible, because music is outside all measure, irreducible to the time spent producing it. The impossibility of comparing two exchange-values on the basis of the labor of the composers and performers announces the impossibility of a differential pricing of music, but also of the impossibility of relegating the production of signs in representation to labor-value. Moreover, the use-value of music is in the spectacle of its operativity, of its capacity for creating community and reconciliation; it is in the imaginary of the simulacrum of sacrifice. This use-value has no relation to the labor of the musician considered separately, since it only has meaning in, and by, the "labor" of the spectator. Thus usage and exchange diverge from the start.

Nevertheless, representation was able to make people believe, for two centuries, that it was meaningful to have a measure for value, that exchange and usage existed and came together in value. Music announced this mystification, made it potentially legible: in representation, music is exchanged for what it is not and is used as a simulacrum of itself. All of the rest of production is also a simulacrum of order in exchange, of harmony.

Exchange and Harmony

All of this could have been heard, for representation doubly implies harmony. First, as spectacle, representation is the creation of an order for the purpose of avoiding violence. It metaphorizes the simulacrum of the sacrificial channeling of violence. It enacts a compromise and an order society desires to believe in, and to make people believe in.

Second, classification presupposes a topology and mathematical model: the mathematics available at that time was necessarily based on theories of the machine in equilibrium, in harmony. Here again, music prefigured the trap into which the major part of political economy was to fall, and where it would remain to the present day.

The way in which music elaborated the concept of harmony and laid the foundation for social representation is fundamental and premonitory. We may hazard the hypothesis that the emplacement of the musical paradigm and its dynamic foreshadowed the mutation that ushered in social representation as a whole. More precisely, it foreshadowed the mutation in exchange, which accompanied representation and affected the entire economy, particularly in the way in which the search for harmony as a substitute for conflict and as simulacrum of the scapegoat would come to dominate it.

Music, from the beginning transected by two conceptions of harmony, one linked to nature, the other to science, was the first field within which the scientific determination of the concept would prevail; political economy would be its final victory. Of course, music has been conceptualized as a science as far back as the day Pythagoras supposedly heard fourths and fifths in the pounding of the blacksmith. But, simulacrum of the sacrifice in its most basic form, of the natural ritualization of the channelization of violence, it was first theorized in its relation with nature. Originally, the idea of harmony was rooted in the idea of order through the endowment of noise with form.

> The order of motion is called "rhythm," while the order of voice (in which the acute and grave tones are blended together) is termed "harmony," and to the combination of these two the name "choristy" is given.[70]

Harmony is thus the operator of a compromise between natural forms of noise, of the emergence of a conflictual order, of a code that gives meaning to noise, of a field in the imaginary and of a limit on violence. Harmony theorizes its usage as a simulacrum of ritual by affirming that it has a pacifying effect: *the less of one it has, the more it must say it has one*. Harmony is in a way the representation of an absolute relation between well-being and order in nature. In China as in Greece, harmony implies a system of measurement, in other words, a system for the scientific, quantified representation of nature. The scale is the incarnation of the harmony between heaven and earth, the isomorphism of all representations: the bridge between the order of the Gods (ritual) and earthly order (the simulacrum). 'Music honors harmony; it spreads spiritual influence and is in conformity with heaven: when the rites and music are clear and complete, heaven and earth fulfill their normal functions.'[71] This explains the fundamental political importance of music as a demonstration that an ideal order, the true image offered by elemental religion, is possible.

> This is why the sages of ancient times, believing that by nature all things move, turn, and tend toward and by means of others of their kind, used Music and encouraged its use, not only to give pleasure to the ears, but principally to moderate or stir up the passions of the soul, and appropriated it for their oracles in order to softly instill and firmly incorporate their doctrine in our minds and, by awakening them, elevate them further.[72]

This conception of natural harmony, an inevitable order in the world, is found as late as Rousseau, who argued in favor of natural, Italian music, against the artificial, contrived music of the French; music, according to him, should be a language, it should be evocative of conversation and thus preserve political order.[73] For Grétry and Villeteau, the model for music is declamation. The link between harmony and representation is here clearly evident: harmony presupposes represented dialogue; it leads to the Opera, the supreme form of the representation by the bourgeoisie of its own order and enactment of the political.

But it was at this same time that modern theorizing about the foundations of harmony was born. The idea was no longer to conceptualize music as a naturally ordered whole, but to impose upon it the reign of reason and the scientific representation of the world: harmonic order is not naturally assured by the existence of God. It has to be constructed by science, willed by man.

Theorizing then became the basis for production. The introduction of bar lines in musical notation, of thoroughbass and equal temperament, made music the representation of a constructed, reasoned order, a consolation for the absence of natural rationality.

Music, in its ambivalence, in its all-embracing hope, is simultaneously heard, reasoned, and constructed. It brings Power, Science, and Technology together. It is a rootedness in the world, an attempt to conceive of human creation as being in conformity with nature: "The word *harmony* sweeps its semantic zone with precision: number, artifact, well-being, language, and world" (Michel Serres). Representation would seize upon this vague concept and make it the linchpin of the the order it implies. In this way, the bourgeoisie of Europe finessed one of its most ingenious ideological productions: creating an aesthetic and theoretical base for its necessary order, *making people believe by shaping what they hear*.

Consequently, by observing music at the end of the eighteenth century, or at latest toward 1850, one could have made a serious prediction about the subsequent evolution of the system and about its limits. To make people believe in order through representation, to enact the social pyramid while masking the alienation it signifies, only retaining only its necessity—such was the entire project of the political economy of the last two centuries. The aesthetics of representation could no longer find acceptance as a natural fact. So it disguised itself as a science, as a universal law of perception, as a constructed system of thought.

Thus in the eighteenth century, reason replaced natural order and appropriated harmony as a tool for power, as proof of the link between well-being and science. To those who availed themselves of it, music made harmony audible. It made people believe in the legitimacy of the existing order: how could an order that brought such wonderful music into the world not be the one desired by God and required by science? The two harmonies, the divine and the scientific, combine in the image of a universe governed by a law both mathematical and musical. A law of gravitation and attraction, the melodies of which were calculated by Kepler himself.

An ideology of scientific harmony thus imposes itself, the mask of a hierarchical organization from which dissonances (conflicts and struggles) are forbidden, unless they are merely marginal and highlight the quality of the channelizing order. This idea also figures in the political economy of the period, and later in the theory of general economic equilibrium: exchange is the locus of order, a means for the channelization of discord. Representation entails exchange, which is legitimate only if it creates harmony. In music, harmony was conceived as conciliator of sounds, an equilibrium in the exchanges of sound matter; in the economy, it was theorized as an equilibrium in exchanges of flows. In political economy, as in music, conciliation was an end in itself, independent of the usage of the flows, represented as abstract and objective quantities. Before political economy, then, music became the bourgeoisie's substitute for religion, the incarnation of an idealized humanity, the image of harmonious, nonconflictual, abstract time that progresses and runs its course, a history that is predictable and controllable. This order is subtle: it does not operate by compelling uniformity, but is on the contrary indissociable from difference and hierarchy. The harmonic system functions through rules and prohibitions: in particular, what is prohibited are repeated dissonances, in other words, critiques of differences, and thus the essential violence. Harmony lives by differences alone, for when they become blurred there is a potential for violence. Difference is the principle of order.

Ulysses' monologue in Shakespeare's *Troilus and Cressida* brings out these relations of necessity between harmony and music, between differences and hierarchy. The text's much-quoted musical metaphor clearly defines order as a system of vibrating chords separated by intangible differential intervals. Without differences, the strong prevail and the weak are crushed; harmony is the hierarchical system that protects the weak, while maintaining the fixed differences distinguishing them from the rich:

> Take but degree away, untune that string,
> And hark what discord follows![74]

In summary, representation leads to exchange and harmony. It requires a system of measurement, an autonomous value for the work, and hierarchy. Even though representation may lead to the enactment of a conflictual classification of social realities (divided, for example, into social "classes"), their representation in the theater of politics inevitably leads to the organization of harmonious exchange, fixed borders, compromise, and equilibrium. No system of representation can find a lasting foundation in the absence of harmony. To make people believe in what is represented in such a system, it is necessary at a certain point to put an end to dissonance, to announce compromises. Representation thus excludes the possibility of a triumph of dissonance, which would be an expression of lack and the failure of the channelization of the imaginary. Entrapping the social form in exchange is just another way of drawing the theoretical debate into the Manichaeism of representation, and at the same time toward compromise. Marxism did not escape this confinement. Trapped within representation, it culminates in speculative abstraction and conflict between social classes—themselves abstract representations of the real—and leads ineluctably to compromise and order.

Harmonic Training

Making people believe in something so contrary to the contradictory reality of society, making musicians who came from the common people into the spokesmen for a harmonic order—this required a fantastically efficient process of normalization, a training process, a marking of the creator and listener alike. The normalization of the musician, for the purpose of turning him into the producer of an order and an aesthetics, was to be the dominant trend of this period. The normalization of music meant first of all the normalization of musicians, performers, and composers, who up to that time had remained undifferentiated. In fact, at the beginning of the period, surveillance and training was very rigorous and very efficient. This memorandum addressed to Bach by the consistory of Leipzig on February 16, 1730, gives something of an idea of the type of control that was exerted over creators of all kinds in the first days of tonal music:

> Whereas attention has been called to the fact that in the public divine services during the past Advent Season the chanting of the Nicene Creed has been omitted and it has been desired to sing and introduce new hymns, hitherto unknown, but such an arbitrary procedure is not to be tolerated. Now therefore . . . we herewith require same that he shall arrange that in the churches of this town, too, matters shall be regulated accordingly, and new hymns, hitherto not customary, shall not be used in public divine services without . . . our previous knowledge and approbation.[75]

A little later, control over musicians' production was assured not through strict control over musical production itself, but through the supervised freedom of the producers. Thus, the conservatories were charged with producing high-quality musicians through very selective training. Beginning in the eighteenth century, they replaced the free training of the jongleurs and minstrels with local apprenticeship, as evidenced by the following notes from a conversation held in 1771 between Burney and Piccini on the subject of the Conservatorio de Sant' Onofrio, near Naples:

> Boys are admitted from eight to ten to twenty years of age . . . when they are taken in young they are bound for eight years. . . . After boys have been in a conservatorio for some years, if no genius is discovered, they are dismissed to make way for others. Some are taken in as pensioners, who pay for their teaching; and others, after having served their time out, are retained to teach the rest. . . . The only vacation in these schools in the whole year is in autumn, and that for a few days only: during the winter, the boys rise two hours before it is light, from which time they continue their exercise, an hour and a half at dinner excepted, till eight o'clock at night. . . . In the common practising there was a "Dutch concert," consisting of seven or eight harpsichords, more than as many violins, and several voices, all performing different things, and in different keys: other boys were writing in the same room; but it being holiday time, many were absent who usually study and practise in this room.[76]

It might be thought that this confinement in the conservatory ended and freedom was attained in the nineteenth century. Nothing of the kind happened. The confinement lasted as long as representation did; to be convinced of that, it is sufficient to read this extract from a report prepared by Charles L'Hôpital, inspector general of public instruction, for the Commission for the Renovation of Musical Studies (1928-31), dated October 24-31, 1931:

> So, do you not think that in our day and age, now that our ideas have become more liberal and less burdened by conventional fears, as the result of the general evolution of minds and morals—not what it presents in the way of excess, but what it offers that is most reasonable—it would be strongly desirable that we combine on more than one occasion our groupings of male and female students for the practice and performance of truly choral works, that is to say *mixed works*, it being well understood that study in detail would be pursued separately. I believe that with prudence and vigilance—excepting, of course, cases where there is a counterindication, and that could happen—these gatherings, the very idea of which would have been a shock to the imagination a mere forty years ago, in our time would quickly gain favor due to the singular intensity of the artistic attraction they would surely exert over the performers and those around them.

Harmonic Combinatorics and Economic Development

Harmonic representation created a code taking combinatorics as the basis of its dynamic and as frame of its expansion. In reality, the only freedom the order of tonal music left open was the freedom to express oneself within its rules. Thus harmony necessarily implies a combinatorics—a system governing the combination of authorized sounds, from the simplest to the most complex, the most abstract, the longest. This combinatory dimension may seem to be absent in political economy; in fact, though it is accurate to say that it is not often found in theoretical discourse itself, it is present implicitly at every turning point in the history of the economic and political sciences of representation.

In 1785, when the first reflections on the possibility of political representation and the electoral system of decision making were beginning to appear, Condorcet demonstrated the cardinal importance of combinatorics in the choice of coherent procedures. The classification of possible combinations, and compromises between them, lay at the heart of the debate over choosing between different voting procedures.

In the theory of representation combinatorics also plays a role in the analysis of conflictual strategies: game theory, the confrontation and arbitration of strategic combinations (discrete or continuous), the foundation for compromise in development. Tuning to the oboe's *la* before beginning to play is a prior compromise made by the musicians in an orchestra, one that is shielded from subsequent challenges by the orchestra leader.

Thus by listening to music, we can interpret the growth of the European economy and the political economy of the eighteenth and nineteenth centuries, not as an incomprehensible and miraculous accumulation of value, but in the context of the idea of *combinatorics*: eighteenth-century science made possible a broader range of combinations of available materials; it allowed the exploration of larger aggregates and their representation in simple terms. All of the theoretical and material dimensions of the field opened were thus accessible, allowing growth by the combination of simple elements. From this point of view, we are led to the hypothesis that the major theoretical discovery of the nineteenth century, from the point of view of economic growth, was the mathematical discourse that made possible a combinatory representation of the quasi totality of functional relations, namely Fourier's serial decompositions of 1800. Once each function is decomposable into a polynomial, each noise becomes decomposable into elemental sounds and recomposable into large-scale aggregates of instruments. Capitalism was to translate this proposition into: "all production is decomposable into a succession of simple operations and recomposable into large-scale factories"; efficiency required, for a time, gigantism and the scientific division of labor.

As in music, combinatorics in production is thus central to the search for and formation of compromises, of harmony between divergent interests. But combinatorics is only possible in the limited field of discrete and controllable sounds. Beyond that, it gives way to statistics, macroeconomics, and probability: before the economy did, music demonstrated that combinatory growth explodes in the aleatory and the statistical. Harmony—order in exchange, the creator of a form of growth—was to produce the conditions for its own undermining. This form of music, then, itself produced the processes that were to destroy its codes when it reached its limits. Once again, music was prophetic; it experienced the limits of the representative mode of production long before they appeared in material production.

Similarly, the emergence of large orchestras and the limitations on their growth would have offered the system, if anyone had cared to listen, a premonitory indication of its evolution: the orchestra—an idealized representation, in the field of the sign, of the harmonic economy and the order implied by the orchestra leader—also reached the outer limits of the functioning of harmonic production.

The Metaphor of the Orchestra

The representation of music is a total spectacle. It also shapes what people see; no part of it is innocent. Each element even fulfills a precise social and symbolic function: to convince people of the rationality of the world and the necessity of its organization. In accordance with the principles of exchange, the orchestra in particular has always been an essential figure of power. A specific place in the Greek theater, it is everywhere a fundamental attribute of the control of music by the masters of the social order. In China, the emperor alone had the right to arrange his musicians in four rows in the form of square; important lords could have theirs on three sides, ministers on two, and ordinary nobles in a single row.[77]

The constitution of the orchestra and its organization are also figures of power in the industrial economy. The musicians—who are anonymous and hierarchically ranked, and in general salaried, productive workers—execute an external algorithm, a "score" [partition], which does what its name implies: it allocates their parts. Some among them have a certain degree of freedom, a certain number of escape routes from anonymity. They are the image of programmed labor in our society. Each of them produces only a part of the whole having no value in itself. For a long time, the leader who directed them was just one of the musicians (Haydn directed on violin or harp), or was one element in the spectacle, as in Indian orchestras:

> He [beats time] by tapping with his fingers on each side of a sort of drum tightly braced. As he beats, his head, shoulders, arms, and every muscle of his frame, are in motion. He rouses the musicians with his voice, and animates them with his gestures.[78]

The orchestra leader did not become necessary and explicit until he was legitimated by the growth in the size of orchestras; he was noise at first. Later he was symbolized in abstract signs, at the culmination of a very long process of the abstraction of regulatory power. Up to and including Beethoven, even symphonies were performed by a small number of musicians (twenty-three for the Ninth),[79] with no leader. But combinatorics entails growth, and growth entails the leader. After 1850, when the size of the audience and the halls made it feasible, the same works were played with over one hundred musicians, with duplication of instruments. Berlioz, the "organizing conductor,"[80] was one of the first to mount the rostrum and, beginning in 1856, gave theoretical expression to this power. In the theory, the orchestra leader appears as the image of the legitimate and rational organizer of a production whose size necessitates a coordinator, but dictates that he not make noise. He is thus the representation of economic power, presumed capable of setting in motion, without conflict, harmoniously, the program of history traced by the composer. The theory elaborated by Berlioz is, moreover, a theory of power; only a few words of the following text would have to be changed to make it pure political theory:

> The obligation on the part of the performers to look at the conductor implies an equal obligation on his part to make himself visible to all of them. Whatever the arrangement of the orchestra may be, whether on steps or on a horizontal plane, the conductor must select his place so that he can be seen by everybody. The greater the number of performers and the larger the space occupied by them, the higher must be his place. His desk should not be too high, lest the board carrying the score hide his face. For his facial expression has much to do with the

influence he exercises. If a conductor practically does not exist for an orchestra unable or unwilling to see him, he exists just as little for one unable to see him completely. Noises caused by striking the desk with the stick or by stamping feet are to be banned completed; they are not only inexpedient, they are barbaric.[81]

This notion of the conductor as a leader of men, simultaneously entrepreneur and State, a physical representation of power in the economic order, has since that time never been absent from the discourse on music. Lavignac and Laurencie's *Encyclopédie de la musique* presented the conductor this way in 1913: "In summary, the orchestra leader must possess the qualities of a leader of men, an always difficult task that is more particularly delicate in the case of artists."[82] Or again, more recently:

And where dictators want robots, captains look for responsive and responsible mariners. The leader is second in command, the strings hoist the sails, the kettledrum beater is the helmsman. (This, by the way, disposes of often attempted experiments in the "conductorless orchestra." Uncommanded crews, if competent, will do well enough in the quiet waters of routine; but on the high seas somebody has to take command, for previous planning of the voyage will be of little avail—or else immoderately careful sailing will make the proudest ship cut a sad figure.) Our hero, we confide, will be of the captain type: a good sailor, a good drinker, and a good curser, and on the whole the most pious—that is, the least megalomaniac—of men.[83]

Thus, the legitimated leader gradually loses his most visible attributes: the baton shrinks and even disappears. The necessity of power no longer needs to be established. Power is; it has no need to impose itself; and the technique of conducting evolves from authority toward discretion.

The ruling class—whether bourgeois industrial or bureaucratic elite—identifies with the orchestra leader, the creator of the order needed to avoid chaos in production. It has eyes only for him. He is the image it wishes to communicate to others and bestow upon itself. But all of the spectators cannot easily identify with him, and his monologue can become unbearable unless representation can provide a wider field of identification. Other figures are then necessary and, I think, one can toy with the idea that this is the essential role of the concerto in representation: the leader is no longer alone; the soloist emerges, the enactment of the individual who has risen above the crowd. Despite his localized task, he can address the leader and the other musicians as a group on an equal footing; he can, in his turn, direct the entire social apparatus, look upon and master the world of which he is a part, play his role and dominate the group as a whole, without, however, aspiring to the ponderous totality of power. He saves the spectator from having to choose between identifying with the anonymity of the musicians and the glory of the leader.

The Geneology of the Star

Once the code was in place, the loudest voices used it to make themselves heard. The combinatory game of music became tied in with economic growth and the contradictions of capitalism, and exchange appeared for what it is: a mask for possession and accumulation. For in introducing music into exchange, representation submitted it to competition. It thus necessarily entailed the triggering of the process of selection and concentration—the durability of those who adapt the best to the system's rules of functioning—and made it impossible to preserve a localized, nonhierarchical usage of music. Selection and universal consumer access to the same music set a process in motion whereby the market expanded for certain musicians, and disappeared for others. In representation, localization quickly became incompatible with exchange. Production for a wide market became the rule, paving the way for mass production, after replication became possible. Thus if the star preceded repetition, both were a consequence of the entry of money into music. Ruthlessly, the logic of political economy accelerated the process of the commodification of music, the selection and isolation of the musician, enriching those who were "profitable"—in other words, the *stars*—and producing a new kind of consumer good, necessarily implied by the very rules of competitive exchange—*success*. The characteristics of success would be dictated by the economy: brevity (reduced labor costs), quick turnover (planned obsolescence), and universality (an extensive market).

The nineteenth century created the technical conditions for this process: the star and copyrights. In the twentieth century, the phonograph record would complete the process and disrupt the network of music. The genealogy of these phenomena is of cardinal importance: the grinding deformation of the social position of the musician, the rerouting of usage toward the spectacle in the interests of exchange. But it is not all that simple; the process of competition alone cannot explain the mysteries of fame. Even if the musician must bend to the rules of capitalist production to be easily heard, he very often knows how to play a double game and make himself heard despite the silent partner. But when he is caught in the act of willful blasphemy, he is destroyed, or else recuperated and travestied.

The genealogy of the star is one of the very first manifestations of the deterritorialization of representation, the disordering of competition in space, and later even in time.

The Genealogy of the Classical Interpreter

The star system, the outcome of competition, began in the middle of the nineteenth century, when a repertory was constituted, in other words, when Liszt, in 1830, began to play the music of other contemporary composers in concert, and Mendelsson played Bach (on the occasion of the centenary of the *Saint*

Matthew Passion in 1829). Liszt gave repertory a spatial dimension and Mendelsson, a temporal dimension. These two dimensions were necessary to the expansion of the music market, for there is no broad outlet without syncretism or universalism. At the same time, this representation confirmed the onset of a fear: society was attempting to rediscover the music of past centuries, sensing perhaps that its own systems of protection against violence and for the channelization of the imaginary were losing their effectiveness. Thus the process of the emergence of the star in classical music was based on the valorization of a stockpile: the market expanded not only through the creation of new products better adapted to needs, but also through an increase in the number of people who could consume old products. As if the proletarians who were gaining access to music hoped to revive the past power of the lords and an ersatz bourgeois spectacle.

Music's mode of financing then completely shifted, making publishers partial substitutes for patrons. Interested in the production of new works, they took the risk of sponsoring them for a rapidly expanding market of amateur interpreters. The bourgeoisie, unable to afford a private orchestra, gave its children pianos. There was a need, therefore, for productions that could be played on them. Works for a small number of instruments, or adaptations of that kind, were thus preferred by publishers.

The breadth of the piano repertory of the nineteenth century is quite clearly connected to the place it occupied in the salons of the bourgeoisie of the time, as an instrument of sociality and an imitation of the Parisian salons and the courts. Power continued to address the musician haughtily. But the tone was no longer one of conquest; it was the tone of the grocer:

By a little management, and without committing myself, I have at last made a complete conquest of that *haughty beauty*, Beethoven. . . . I agreed with him to take in MS. three Quartets, a Symphony [the Fourth], an Overture [*Coriolan*], a Concerto for the Violin, which is beautiful, and which, at my request, he will adapt for the pianoforte, with and without additional keys; and a Concerto for the Pianoforte, for *all* which we are to pay him two hundred pounds sterling. The property, however, is only for the British Dominions. Today sets off a courier for London through Russia, and he will bring over to you two or three of the mentioned articles. Remember that the Violin Concerto he will adapt himself and send it as soon as he can. The Quartets, etc. you may get Cramer or some other very clever fellow to adapt for the pianoforte. The Symphony and the Overture are wonderfully fine, so that I think I have made a very good bargain. What do you think? I have likewise engaged him to compose two sonatas and a fantasia for the pianoforte, which he is to deliver to our house for sixty pounds sterling (mind I have treated for Pounds and not Guineas). In short, he has promised to treat with no one but me for the British Dominions.

In proportion as you receive his compositions you are to remit him the money; that is, he considers the whole as consisting of six articles, viz. three quartets, symphony, overture, pianoforte concerto, violin concerto, and the adaptation of said concerto, for which he is to receive £200.[84]

No one experienced the insecurity of the musician-entrepreneur more intensely than Mozart, a victim of ruthless economic censorship his entire life, one of the first prisoners of abstract, anonymous money, blackcoat money.

Just a few months before his death, he was still writing letters like this:

Instead of paying my debts I am asking for more money! . . . Owing to my unfortunate illness I have been prevented from earning anything. But I must mention that in spite of my wretched condition, I decided to give subscription concerts at home in order to be able to meet at least my present great and frequent expenses, for I was absolutely convinced of your friendly assistance. But even this has failed. Unfortunately fate is so much against me, *though only in Vienna*, that even when I want to, I cannot make any money. . . . A fortnight ago I sent round a list for subscribers and so far the only name on it is that of the Baron van Swieten![85]

Mozart died several months later. His fortune amounted to 60 florins, according to Constance.[86] He was 3,000 florins in debt (1,000 of which was owed to Puchberg). Baron van Swieten paid for the least costly burial possible. Constance, before remarrying, succeeded in selling eight manuscripts to the king of Prussia for 800 ducats.

It thus became almost impossible to have one's music heard without first being profitable, in other words, without writing commercial works known to the bourgeoisie. To be successful, a musician first had to attract an audience as an interpreter: representation takes precedence over composition and conditions it. The only authorized composers were successful interpreters of the works of others. The spectacular, the exploit, took center stage. It was necessary to sell oneself to have the right to create. Chopin, who, unlike Liszt, made his living from his lessons far more than from his rare concerts, analyzed this process very lucidly:

Today, having lost all such hope, I have to think of clearing a path for myself in the world as a pianist, and I must put off until later those higher artistic aims your letter so rightly presents. . . . There are many talented young men, pupils of the Paris Conservatoire, who are waiting with folded hands for the performance of their operas, symphonies, cantatas, which thus far only Cherubini and Lesueur have seen on paper. . . . In Germany I am known as a pianist; certain music journals have mentioned my concerts, expressing the hope that I may soon take my place among the foremost virtuosi of that instrument. . . . Today I find an unequalled opportunity to fulfil the promise innate in me; why should I not seize it? . . . Ries found it easier to obtain laurels for his *Braut* [*Bride*] in Berlin and Frankfurt because he was known as a pianist. How long was Spohr known only as a violinist before he wrote *Jessonda, Faust*, and so on?[87]

In the same period, Wagner expressed his irritation at the growing status of interpreters:

Every musical composition has had to resign itself, in order to win the approbation of the public, to serving as the instrument and pretext for the capricious experiments of the performers. . . . The musician who today wishes to win the sympathy of the masses is forced to take as his point of departure this intractable pride of the virtuosos, and to reconcile the miracles expected from his genius with such servitude. . . . It is particularly in the singing profession that the abuse we are drawing to attention has built a pernicious empire.[88]

"I play the piano very well," wrote Bizet to an unidentified Belgian composer, "and I make a paltry living at it, because nothing in the world would convince me to be heard by the public. I find this trade of the performer odious! Yet another ridiculous aversion that costs me on the order of fifteen thousand francs a year. I play now and then at Princess Mathilde's and in a few homes where artists are friends and not employees."[89]

The nineteenth century was the period when concerts were at their height. *Le Ménéstrel*, a musical journal of the time, was constantly repeating that it was impossible to print a complete listing of the multitude of concerts taking place. Finding work was thus not a problem for musicians, which explains their subsequent opposition to the phonograph record. The virtuosos (Paganini, Gottschalk, Liszt) commanded a considerable fee, usually fixed, but sometimes proportional to receipts, at least in part. These stars (*vedettes*) worked hard: in 1844, Liszt gave six concerts in fifteen days in Lyons. Gottschalk gave between seventy and eighty concerts a year.

But in time the relation between the interpreter and the work would change. At the time Chopin was composing, the number of professional interpreters was still small, and their market restricted; traveling was time-consuming and a tour gave the artist free time to write and get to know his audience. The audience, which would never hear more than one or two interpretations of a work, did not have standardized criteria for choosing. Today, the process has evolved considerably. Musical representation has made a selection from the huge stockpile passed down from earlier centuries, a stockpile to which additions are still being made, but only to a very slight extent. Musicians who were very well known in their time have disappeared. Others less renowned have survived. Still others reappear with the changing rhythm of fashion. Some interpreters have played the role of a memory—transmitting interpretation and technique, intermediaries of music, actors in a play they did not write, guardians of music—until the mutation of recording, repetition.

Actually, the definitive birth of the star took place when popular music entered the field of the commodity. The evolution of the star is what really developed the economy of representation and necessitated a guarantee of remuneration, an exchange-value, for popular production, which had been overlooked by the creators of the copyright.

The Genealogy of the Popular Star

The process of the selection and emergence of stars in the popular song of the eighteenth and nineteenth centuries relates to same dynamic of musical, cultural, and economic centralization. Up until that time, popular song found expression mainly in the street, the traditional domain of the jongleurs. Its confinement and pricing, first in the cabarets, then in the café concerts, was the precondition for its entry into the commodity market and competition. In the middle of nineteenth century, these halls became the heart of the economy of music; they were an essential source first of exchange, then profit, and gradually replaced the other sites of musical expression, whose capacity to realize surplus-value was insufficient.

The singer-musician gradually ceased to be also an acrobat. The division of labor did its work and, beginning in the seventeenth century, and particularly in the eighteenth, the two professions became completely separate. Acrobatics was confined to the circus, as were practically all spectacles of the body. The popular musician sought other outlets for his work. Music publishing seemed to offer one. However, prior to the eighteenth century, popular music was the object of very few published editions, due to the small market for it. The only published editions that existed were those for distribution by street hawkers, the only possible channel of distribution among the people, whose right to assembly was severely limited. In addition, pirated editions were only suppressed if they were of "dramatic works," not songs.

Musical invention was thus practically nonexistent, and what little there was consisted essentially of texts to be sung to a few *timbres*, or popular tunes, which were engraved in the social memory like a residue of musical ritualization. In 1811, in the journal *La Clé du Caveau* (*The Key to the Cabaret*) we find a list of *timbres* numbering 2,350, but most were rarely used.

Neither the law of 1791 nor the Penal Code of 1810 protected these works, considered petty and unworthy of protection. This form of expression was still strictly controlled in order to prevent songs from becoming vehicles of subversion. Publisher's privileges prevented singers "from having anything published in their name"; "songs could only be printed by a bookseller-publisher." But the police were never able to ensure the enforcement of the law, and numerous clandestinely published songs were circulated: the monopoly over music was one of the first destroyed by the people, before they tackled the others. In order to control these publications, so dangerous for order and the economy, certain eighteenth century publishers proposed to give official status to street singers, as had been done with book hawkers. This would have allowed the selection of a certain number of duly authorized singers, and then the establishment of corporatism, the control of those unable to enter the capitalist mode of production. In the absence of such a status, the police early began keeping these singers under surveillance. Noise-making, subversive, they peddled news that power wished to hide from the people. In a police report of 1751, we read:

> Most disreputable people, like beggars and women of ill repute, when they meddle in singing, do contortions in the streets, and, sometimes sodden with drink, expose themselves to the ridicule of the public, and often add things that are not in the song.[90]

In the nineteenth century these musicians would be gradually driven off. It would only be possible to make music in a fixed-price performance, in other words, in a concert hall. All arguments were valid in the effort to destroy these singers. E. Fétis, director of the *Revue musicale*, which played an important role in the taming of popular music, wrote in 1835:

> Under the government of the Restoration, organ players were accused of having shameful ties to the political police of the kingdom; it was claimed that they were paid to station themselves in front of places the authorities wanted to spy on.[91]

This article provides interesting information on the situation of street musicians at the beginning of the nineteenth century. The police found it hard to tolerate them, and they were increasingly barred from the courts on the pretense that thefts had been committed. The repertory was meager, and the tunes were given little diversity (the tune of "Cadet Roussel" or "Fanchon" ["Kerchief"]), to make it possible for the musically illiterate public to commit them to memory. The singers would sell the lyric sheet for ten centimes. Popular songs were thus extremely impoverished musically. Among the instruments in use that Fétis mentions, the most common were the barrel organ, more or less well tuned, mechanical organs with miniature figures dancing on a little scene, and the bird organ, or *serinette*. The mechanical nature of these instruments is an indication of how limited was the musical repertory in use. The clarinet was essentially the instrument of the blind (with their little dogs). There were in addition a few flageolet players. The violin was high on the list, followed by a small number of harps. Fétis remarks that what he calls the "Aristocracy of the street singers,'' the Italians who accompanied themselves on the guitar and sang *cavatinas*, was little appreciated by the public. He concludes:

> The government could greatly improve the street music of Paris and exert a powerful influence on the direction of the moral pleasures ambulant musicians procure for the people. This is its duty. For a very modest recompense, it would have in its pay a considerable number of musicians equipped with always well-tuned instruments, who would only play good music. The singers, blessed with manly and virile voices, would sing for the people exclusively patriotic hymns and songs whose lyrics, sternly chaste, celebrate the noble virtues and generous actions for which the people have a natural feeling. Instead of singing about being drunk on wine and the pleasures of the brute passions, the people would hear praise for the love of labor, sobriety, economy, charity, and above all the love of humanity.

That is what they wanted to do to popular music in 1835. It says everything there is to be said: about aesthetics and political control, about the rerouting of popular music toward the imposition of social norms.

In 1834, a law was passed that organized the profession and realized the project of the eighteenth-century publishers; it required street singers to wear a badge, as a way to keep tabs on them and limit their number. A little later, after the publishers and songwriters had won recognition of their ownership rights over the sale and performance of their songs, they used the street singers as door-to-door salesmen, as sales representatives: they no longer sold simple texts, but "small format" books, popular publications containing the lyrics and the vocal score. Melody was then able to diversify, and representation was no longer confined to the festival, but was already becoming a sales tactic for an immaterial object, which is what it would be entirely after the appearance of the phonograph record. At noontime in the factories the workers, score in hand, would sing under the direction of a street musician. This became the favorite distraction of the dressmaker's apprentices, whose virtues so many verses celebrate, reflecting the market.[92] Thus there was already a link between the songs of the street and people buying their own idealized image. The "small format" book, the foundation of commodity exchange in localized representation, remained for a long time the sole commercial product of popular music. An entire industry followed in its wake. In 1891, the magazine *Gil Blas Illustré* was founded; each weekly issue contained the text and melody line of an illustrated song.

But street singing did not permit the evolution of the star or an extensive market: the shifting site of the performances prevented a stable market for competing singers from developing. The innovation that truly led to the birth of the music industry was the confinement of popular music in the café concerts and cabarets. Beginning in 1813, on the twentieth of each month, *hommes d'esprit* ("men of spirit") would meet at the Caveau on rue de Montorgueil. During the meal, each person had to sing a song of his own composition. The owner of the cabaret, M. Baleine, rented private rooms to members of the bourgeoisie who wanted to hear these *chansonniers* ("singer-songwriters"). Such associations between singers and cabaret owners multiplied and quickly reached the bourgeois audience. Alongside the café concerts, the *goguettes* developed, in other words, associations of working-class poets, the chansonniers of the people. We find them in particular in Paris and its suburbs. (La Ménagerie, Les Infernaux, Les Bons Enfants, La Camaraderie . . . [The Menagerie, The Infernal Ones, The Good-Natured, Camaraderie].) But while the satire of the Caveau was harmless, the songs of the *goguettes* quickly came under fire. For example, C. Gille of "La Ménagerie" served six months in prison under Louis-Philippe, and his comrades fell at the barricades in 1848. The songs created by the *goguettes* would afterwards serve as an important political catalyst, spread the length and breadth of France by street hawkers. One of the best known *goguettes*, "La Lice Chansonnière" ("The Singer's Arena") stayed open under the Second Empire, uninterruptedly displaying the republican insignia above its entrance.

Since the right of assembly was restricted to solvent consumers, new modes for the distribution of popular music were instituted. The lawmakers of the Second Empire, understanding this danger, banned certain of these forms of representation under the pretense of regulating the right of assembly. Beginning in this period, both the bourgeoisie and the workers listened to songs in entertainment halls, café concerts, and cabarets. Song, which until that time had occupied an essential place in private life, henceforth became a spectacle. Representation set itself up in opposition to lived experience—as had learned music, though a century earlier. Popular music did it in a much more flexible manner, however: silence was not the rule. The representation of popular music suppressed neither festival nor the threat of subversion. The café concert, where the singer was paid as such, was born in 1846, when the "Café des Aveugles" ("Cafe of the Blind") presented the first "concerts" of popular music. Standing on a platform supported by two barrels, the singers received no recompense from the management of the establishment: they passed the hat, sold copies of their songs, and sang by the tables. Representation thus remained interwoven with life. If the author was present, he collected his royalties directly. If not, there was no guarantee. The café concert, originally intended for the bourgeoisie, under the Second Empire became a popular festival, transformed into the *caf' conç'* (in *Viens poupoule* [Come, My Darling], Saturday evening after the work grind, the Parisian worker says to his wife for a treat, "I'll take you to the café concert").[93] The *caf' conç'* vogue was due in part to the talent of the artists and composers, but a large part of it had to do with the unrestricted atmosphere (smoking and drinking were allowed) and how cheap the admission fee was, when there was one at all.

Confinement was channeled into festival, and representation remained the locus and pretext for the circulation of ideas. After 1850, the *caf' conç'* began to multiply. By 1870 there were more than 100 in Paris, and the trap of commercial selection closed tightly on popular music. The chansonniers adapted their texts to their audience, and roles were defined. There was success for comics playing dotards (Polin, Dranem, and at the beginning of his career Maurice Chevalier . . .), and for melodramatic songs "in which the unwed mother, the corrupt bourgeois, the sacrificed soldier, the honest worker, the sailor at sea, the infamous seducer and the drunk, in turn tyrant and victim, were central figures."[94] From 1900 on, singers fell into stereotypes (the off-color comic, the aging lady's man, the singer with a memorable voice) tailored to the audience's tastes. Political songs were also to be heard: the patriotic songs of the Franco-German War of 1870 and of 1886-87 were replaced, during the great economic crisis of 1890-1910, by socialist or openly anarchist verses.

In contrast, the *cabaret*, a variety of the *caf' conç'*, attempted to organize the commercialization of high-quality songs. In 1881, Rodolphe Salis' cabaret, the "Chat Noir" ("Black Cat") was founded; Maurice Donay and Jean Richepin performed there. It was at the cabaret also that the songs of Pottier (whose works were not published until the year he died, 1887), J.-B. Clément, and Nadeau (without a doubt the greatest songwriter of the nineteenth century) were sung. "Temps des cerises" ("Cherry Time") first appeared in a cabaret, later to be circulated in the *caf' conç'*. A great many cabaret songs later became *caf' conç* successes. The clientele was not the same: that of the *caf' conç* was the common people, while that of the cabaret was "bohemian," student, or bourgeois. The bourgeoisie also directed much more criticism toward the *caf' conç*, the "place of debauchery" of the common people, than toward the cabaret: "Song properly speaking, the repository of the age-old essential verve of our national character, popular song, the expression of the informal genius of France, has fallen, in the café concerts, to untold depths of roguish stupidity. The finest lips have lent their audacious grace to this degeneracy, which would be consumed today, in spite of the venerable Caveau, if you and your friends, the poet chansonniers of Montmartre and the 'Chat Noir,' had not triumphantly restored that Parisian gaiety which shines afar," wrote Sully Prudhomme in his epistolary preface to Maurice Boukay's *Nouvelles chansons (New Songs)* published under his real name, M. Couyba. Boukay's "red songs" did not prevent him from becoming minister of commerce under Caillaux in 1911 (demonstrating, as though it were still necessary, the institutional ties between music and money).

The star emerged with the *caf' conç*, thanks to tours by musicians that created an expanded market in the provinces. The first star was undoubtedly Theresa, toward 1865 (with her "bearded lady"). The *tube*, or hit song, also dates from this period, when it was called the *scie* ("saw"). The stars began to garner large fees, and it suddenly became possible to make a fortune through the practice of popular art. Paulus acquired a huge fortune before sinking into poverty. This wealth was strongly criticized by the industrial bourgeoisie and the intellectual elite, who were scandalized by such earnings, and only frequented the wings and recesses of the *caf' conç*. For the working world, on the other hand, these earnings meant dreams of social advancement. Social climbing and the star system have been profoundly interconnected ever since: the image of the group member who made it is a formidable instrument of social order, of hope and submission simultaneously, of initiative and resignation.

There developed around the star an entire array of professions (manager, stagehand, administrator), including the claque, whose role was very important: when representation emerged as a new form of the relation to music, it also became necessary to produce the demand for it, to train the spectator, to teach him his role. That is what the claque was for. It would disappear only after the public's education had been completed, and the demand for representation was well established.

The Economy of Representation

If commodity representation emerged in eighteenth-century England, it was in nineteenth-century France that the regulatory process for its generalized commercialization was organized. The right of ownership over representation was until that time reserved for the musicians of the courts and salons. Outside of "dramatic works," performances brought no reward for the authors. Authors only received pay as performers or from the direct sale of their scores. The idea of remuneration for the representation of their works had difficulty gaining acceptance. The reasons for this were, first, given the absence of a popular market, such representations remained very rare; second, it was impossible to keep track of street performances of popular music; and finally, for a long time judges were reluctant to call something that consisted only of a melody line with changing words a musical "work." It was not until the confinement of popular music that a market developed and that authors finally created an institution capable of representing them and winning them compensation. The *caf' conç'* gave this kind of music an exchange-value; this was later recognized by judges, and only afterward by legislators.

This anecdote clearly reveals a fracture, the birth of the popular consumption of commodity leisure: E. Bourget, P. Henrion, and V. Parizot attended a show at the "Caf' Conç' des Ambassadeurs" ("Café Concert of the Ambassadors"), where they heard a song written by the first and a sketch by the other two. After the performance, they refused to pay their check, claiming that the law of 1791 applied to these "works": "You use our labor without paying us for it, so there's no reason why we should pay for your service." The magistrate's court agreed, in a decision of August 3, 1848; that decision was upheld by the court of appeals on March 26, 1849; and the law of 1791 was applied to all musical works.[95]

On February 11, 1850, the same three men, in association with the publisher, Jules Colombier, who had assumed the court costs, founded the Union of Authors, Composers, and Music Publishers (Syndicat des Auteurs, Compositeurs, et Editeurs de Musique: SACEM), the first institution of its kind anywhere in the world. Its function was to demand, on behalf of the authors and editors, payment of royalties for every representation of a musical work, regardless of its importance.

This was perceived by the bourgeoisie as an attack on its privileges: formerly, the bourgeoisie alone had had the right to have financial dealings in music. Money was its kingdom. The common people were not supposed to have anything but street music, music that was "valueless." The music press, moreover, did nothing to make the new institution better known. The reactions ranged from the silence of the *Revue musicale* to the contempt of *La France Musicale*, expressed in the following words:

> Here is what's new. There was just formed an agency for the collection of royalties for authors, composers, and musical publishers. It was M. P. Heinrichs [*sic*] who invented this new gambit, the aim of which is quite simply to collect or help in the collection of royalties from ballads, ariettas, light songs, and potpourris used in salons and concerts. So from now on, one will not be able to sing a ballad without the threat of being taken by the collar on charges of violating private property. . . . How can serious men spend their time on such twaddle? Really! At a time when we must loudly proclaim the freedom of thought, when art must enter the hearts of the masses through dedication, and most especially selflessness, they go bringing up an issue that is as childish as it is ridiculous! Taxing baladeers. . . . Truly, the lack of common sense has never been pushed to such an extreme. If this project is pursued, we will fight it until it is reduced to nothingness. If you create operas, symphonies, in a word, works that make a mark, then royalties shall be yours; but taxing light songs and ballads, that is the height of absurdity![96]

SACEM gave a *value*, in the bourgeois sense of the term, to the music of the people. According to some accounts, Napoleon III allowed the creation of SACEM to thank songwriters for the help they gave him. This interpretation seems to me very hard to accept, when it is well known that during the first years of his reign he imposed a very restrictive law on performances, and the café concerts remained republican for a very long time. In fact, the creation of SACEM seemed harmless (it occupied itself with light songs), and in the spirit of contemporary French capitalism, it helped guarantee respect for the property rights of all.

Thus a real economic market developed for musical performance. It was a market that would create musical works, because the publishers, who now had a direct interest in advancing the musical representation of a work, became promoters. They encouraged the financing and training of performers in order to make a song profitable by having it represented. They even created "courses" in which a pianist in residence would teach new songs to performers, who would then promote them.

France was thus one of the first countries to guarantee copyright protection covering both the written reproduction and the performance of all forms of music. After the Revolution, this right, first recognized by the judges who were the most closely in touch with economic evolution, was later recognized in the law: "The author of a work of the mind enjoys ownership rights over that work *by the simple fact that it was he who created it.*"[97]

Today, any public representation, in other words, one that is not free and takes place outside the family circle, is illegal without the consent of the author or his representative. Once authorized, it earns the author a payment independent of other expenses (the performer's salary, the publisher's fee, taxes), even in the case of a show that is free or loses money.

In the United States, where capitalism took on different forms, the author's ownership of a copyright on performances of his work has not been established as such, and the musician has remained in a weaker position in the face of capital. On all points, copyright protection is not as strong as in France. The Copyright Office fulfills some of the functions of SACEM in protecting property, but a number of competing businesses assure the valorization of the patrimony by picking up reproduction rights. Performances are only subject to the payment of royalties if an admission fee is charged and it is done for profit, which opens the way for all kinds of dodges, since the defining of what is "for profit" can be quite delicate. In addition, when an author works under contract, he loses ownership in favor of the party who requested the work, who is afterward free to have it reproduced as he sees fit.

In the Soviet Union, protection of authors is very weak: royalties are low, and when the "interests of the nation justify it" none at all are paid. The copyright laws were slightly modified by the 1974 ratification of the Universal Copyright Convention, which liberalized Soviet inheritance rights, since it gives heirs post mortem ownership rights for a period of twenty-five years.

The economy of musical representation depends on the effectiveness of authors' associations in detecting instances of representation. No author can alone enforce the payment for his production, because of the multiplicity of possible performance sites. Conversely, no provider of entertainment can deal directly with all of the authors and composers whose repertories he may use in orchestra performances or on records. Author's associations are thus a substitute necessary to the market in which these transactions take place. They are charged with collecting royalties for the public representation of works on their lists (in France, around three million). They pursue their function in relation to entertainment halls, and radio and television broadcast organizations, as well as dances, jukeboxes, stores, fairs . . . The area they cover is vast (there are over 180,000 dances held in France each year), and an institution of this nature requires enormous oversight powers. Each representation is subject to prior authorization. SACEM has at its disposal a listing of all concert halls, which are required to submit their programs; monitoring them is relatively easy. Every orchestra, then, has to furnish a list of the works it performs. The payments made depend on the nature of the establishment. Establishments in which music plays an essential role pay a levy proportional to their receipts. Those in which music plays a secondary role pay a fixed rate set according to the real volume of music usage (square footage, size of clientele . . .). The amount paid is independent of the fame of the author and the quality of the work. In France, the allotment of the anticipated royalties is theoretically one-third for the composer, one-third for the lyricist, and one-third for the publisher.[98]

In no country is the author put in the position of wage earner. There have been several attempts to create associations for which salaried musicians would work. These associations would buy musicians' works at a fixed rate and then try to exploit them. Such an arrangement presupposes that it is possible for the enterprise to sell the music, in other words, to promote it successfully and collect the royalties. But the consumers of music are very dispersed, and it is not within the power of a single enterprise to keep track of them. Thus for the production of representation to be profitable, there would have to be a very substantial infrastructure that representation alone does not justify. Therefore, all of these projects have ended in failure. The only case of salaried musicians is that of composers of music for films, who often receive upon completion of the work a fixed sum that is considered an advance on future receipts, though the royalty rate is so low that the advance is rarely exceeded. All in all, since those who reproduce musical works have an interest in ensuring that the authors receive appropriate compensation, when a work does succeed in attracting an audience, capitalism grants the creator the legal fiction of ownership over his work and assures him an often considerable reward for its use.

Thus in the countries in which the author is best protected, he is in the position of the holder of an estate who entrusts it to a specialized enterprise and delegates all of his rights to it in order to draw revenue. In the economy of representation, profit is linked to the ability of an innovation to accrue value; the remuneration of the author of the innovation can then be a function of the number of valorizations of his innovation, in other words, of the number of representations of his work. Processes such as this exist in many sectors in which it is still possible to identify the creator, the *molder*. For example, in the early days, automobiles were marked with the signatures of the manufacturers and designers, who produced models in limited numbers and were paid in accordance with copyright principles. The author occupies a position upstream of the capitalist process, and his labor is remunerated in the form of a rent.

The commodity quickly became an object of spectacle. Already in the eighteenth century, music-turned-commodity was announcing the future role of all commodities under representation: a spectacle in front of silent people. In representation, commodities speak on behalf of those who purchase the spectacle of their order, their glory. Usage, as soon as it is represented, is destroyed by exchange. The spectacle emerged in the eighteenth century, and, as music will show us later on, it is now perhaps an obsolete form of capitalism: for the economy of representation has been replaced by that of repetition, and in Carnival's quarrel with Lent, it is Lent that has taken the upper hand.

The Drift toward Repetition

The Rupture of Combinatorics: Antiharmony

Harmonic combinatorics and the individualist system of representation necessarily led to a Romantic exacerbation of individualism and a rupture in the process of representation, impelling musicians toward an increasingly clear awareness of their relations with the world, of the divergences between creation and reality. Their music signified lack and organized their own solitude. They no longer vibrated in a world over which they had control, but in a reality foreign to their visions. They were the first to become conscious of the impossibility of definitively establishing oneself in harmony and within the constraints of combinatorics. It was no longer possible for the musician to create within this thoroughly explored code, even if it was still possible to sell the by-products. Schoenberg wrote his first quartets while earning his living by orchestrating operettas. Mahler and Satie were writing at the time the first May Day marches in Vienna took place, when Picasso created a scandal with *Les demoiselles d'Avignon (The Maidens of Avignon)* and Der Blaue Reiter published *A Call for the Emancipation of Dissonance.* The rupture of harmony's relation of dominance was the beginning of the end of the representative network and the mystical fusion of the middle classes with the social order.

In fact, music at the end of the nineteenth century was highly predictive of the essentials of the ruptures to come. And practically everything that happened took place in Vienna: it was there that music announced a decline, a rupture, and simultaneously a tremendous theoretical accomplishment. Musical creation rose to a fevered pitch, exploding prior to the political discontinuity for which it itself, to a certain extent, prepared the way. The present economic crisis and efflorescence of our decadence were preprogrammed in Viennese music. Wagner, by ignoring the simple melody line, was already moving away from the representation of harmony. Then Mahler confirmed the end of an age, giving expression to the dissolution of tonality and the fundamental utopia of liberation, the integration of noise into musical organization, the profound but radically new reinsertion of musical labor into the lives of all men. The musical debates of the turn of the century, then, express the desacralization of musical matter, the advent of the nonformal, the noninstituted, the nonrepresentative. Vienna, where all of this was written, heard, and said; the prewar Vienna of the dodeca-phonic turning point, about 1910, gripped by a self-destructive fascination, in which the Jewish bourgeoisie, by virtue of its multiple belongings and sense of transcendence, would take art to the limits of its potential; Red Vienna, where the street in revolt would attempt the only organization ever to initiate the self-management of concerts, with Schoenberg's "Society for Private Musical Performances."[99] Stefan Zweig, in *The World of Yesterday*, memoirs written in the darkest hours of his expatriation, provides an admirable description of this typically Viennese process:

> Immeasurable is the part in Viennese culture the Jewish bourgeoisie took, by their cooperation and promotion. They were the real audience, they filled the theaters and the concerts, they bought the books and the pictures, they visited the exhibitions, and with their more mobile understanding, little hampered by tradition, they were the exponents and champions of all that was new. Practically all the great art collections of the nineteenth century were formed by them, nearly all the artistic attempts were made possible only by them; without the ceaseless stimulating interest of the Jewish bourgeoisie, Vienna, thanks to the indolence of the court, the aristocracy, and the Christian millionaires, who preferred to maintain racing stables and hunts to fostering art, would have remained behind Berlin in the realm of art as Austria remained behind the German Reich in political matters. Whoever wished to put through something in Vienna, or came to Vienna as a guest from abroad and sought appreciation as well as an audience, was dependent on the Jewish bourgeoisie. When a single attempt was made in the anti-semitic period to create a so-called "national" theater, neither authors, nor actors, nor a public was forthcoming; after a few months the "national" theater collapsed miserably, and it was by this example that it became apparent for the first time that nine-tenths of what the world celebrated as Viennese culture in the nineteenth century was promoted, nourished, or even created by Viennese Jewry.[100]

Political marginality formed the foundation, the infrastructure, of cultural marginality. These two marginalities designated the only two forces that were to survive the destruction of the strength of Vienna: a move toward utopian socialism and a shattering of the constraints on music—as if the cultural powerlessness of Viennese political society to assume its music tolled confirmation of the political death of the entire society of representation. The ruling class, incapable of inventing a music and financing it, would prove incapable of organizing its own economic defense and political survival.

Thus music forced a break with tonality before economic accumulation forced a break with the laws of the economy of representation. Harmony—the repressive principle of the real—after having created romanticism—the utopian principle of the real, the exaltation of death in art—became the death of art and destroyed the real. An excess of order (harmonic) entails pseudodisorder (serial). Antiharmony is the rupture of combinatory growth, noise. At the end of meaning, it sets in place the aleatory, the meaningless, that is to say, as we shall see, repetition.

The lesson taught by music is thus essential and premonitory: with the end of representation, a first phase in the deritualization of music, in the degradation of value and the establishment of political order, comes to a close. Ritual murder recedes behind the spectacle of music. But after the strategy of bestowing form has tried everything, the represented ritual disappears beneath an acceptance of nonsense and a search for a new code. Representation, the substitute for reconciliatory sociality, fails; the rupture of harmony seems to announce that the representation of society cannot induce a real socialization, but leads to a more powerful, less signifying organization of nonsense. If this hypothesis holds true, then modernity is not the major rupture in the systems for the channelization of violence, the imaginary, and subversion that so many anachronistic thinkers would like to see. Not a major rupture, but sadly, boringly, a simple rearrangement of power, a tactical fracture, the institution of a new and obscure technocratic justification of power in organizations.

The probabilist transcending of combinatorics by a code of dissonance, founded on a new mode of knowledge, then announced the advent of a power establishing, on the basis of a technocratic language, a more efficient channelization of the productions of the imaginary and forming the elements of a code of cybernetic repetition, a society without signification—a repetitive society.

Music, exploring in this way the totality of sound matter, has today followed this its path to the end, to the point of the suicide of form. As Jean Baudrillard writes: "In every spectacle (of gigantism), there is the imminence of catastrophe."

The Socialization of Music

Music, seen from the point of view of its codes, heralds a rupture of the representation of harmony, and its political economy is exemplary of that rupture. Three things foster the expectation that representation is becoming an anachronistic form of musical expression incompatible with the requirements of the capitalist economy:

The production of representation has a fixed productivity level, so its costs go up as the productivity of the rest of the economy improves. In itself, therefore, the activity of performing cannot be profitable, and capitalists will stop investing in it.

The love of music, a desire increasingly trapped in the consumption of music for listening, cannot find in performance what the phonograph record provides: the possibility of saving, of stockpiling at home, and destroying at pleasure. We must add to that the fact that the disorganization of urban life makes attending any performance an expedition. Representation can maintain itself only by extending its market (a project undertaken beginning in 1920, the date of the first radio-broadcast concert), by making its production multinational (through the coproduction of operas and concerts). In particular, as we will see later, performance becomes the *showcase for the phonograph record, a support for the promotion of repetition.*

Finally, radio made representation free. A radio station is not a concert hall; it neither pays for the performance nor pays the musicians, on the grounds that the broadcast of a work, live or recorded, gives the work free publicity and is thus advantageous for other forms of the commercialization of music. Not recognizing themselves as a locus of representation, radio stations everywhere wished to be exempted from copyright restrictions and from paying royalties on the objects they use. But the situation varies by country: in France, State radio and television pays the music writers' associations a percentage of the gross profit, minus taxes, earned from broadcast royalties and advertising receipts. Commercial stations pay SACEM royalties proportional to their advertising receipts. In the United States, until the law of 1976, radio stations had succeeded in avoiding payment of royalties to music publishers and record companies. Before 1976, every proposal to impose levies lost under pressure from the radio and television lobby, which is very important at election time. Similarly, jukeboxes, which in the United States are controlled by underworld elements, were exempt until 1976. In the Soviet Union, radio stations do not pay royalties either. Music is thus being remunerated more and more globally, independently of the individual work. But then it becomes impossible to identify the author of the representation. Today, the problem has become almost insoluble: how can authors be remunerated on the basis of the number of representations of their works when the channels and number of representations have been multiplied to such an extent—if not by statistical and aleatory means?

The economic rights and rules invented by competitive capitalism do not apply to today's capitalism of mass production, of repetition. Inevitably, the statistical evaluation of the quantity of the representation will be adopted. The usage of music will be evaluated exclusively by polls determining the quantity of the music broadcast. Musicians will be remunerated according to statistical keys and treated as producers of a stockpile of undifferentiated raw material. This shift relates to a statistical reality: the disappearance of use-value in mass production and the final triumph of exchange-value.

Representation as the Showcase of Repetition

The advent of recording thoroughly shattered representation. First produced as a way of preserving its trace, it instead replaced it as the driving force of the economy of music. Since then, representation survives when it is useful in record promotion or among artists who do not command a significant record audience. For those trapped by the record, public performance becomes a simulacrum of the record: an audience generally familiar with the artist's recordings attends to hear their live replication. What irony: people originally intended to use the record to preserve the performance, and today the performance is only successful as a simulacrum of the record. For popular music, this has meant the gradual death of the small bands, who have been reduced to faithful imitations of recording stars. For the classical repertory, it means the danger (to be discussed later) of imposing all of the aesthetic criteria of repetition—made of rigor and cold calculation—upon representation. Thus, the simulacrum of usage is only retained when it furthers exchange or mimics it. Representation has become a showcase and mimics repetition.

Recording is therefore more than a simple mutation in the technological conditions of music listening. It is also a very deep transformation of the relation to music.

Of course, the mass repetition of the music object leans very heavily upon representation and draws the major portion of its sound matter from it. *Repetition began as the by-product of representation. Representation has become an auxiliary of repetition.* This general phenomenon extends far beyond music itself: the service acts as the showcase and support for the commodity object, after having contributed its structures and inspired it. For example, haute couture, long the model for ready-made clothes, now draws its inspiration from them.

However, when the process of representation is transformed into repetition, there develops a refusal to submit to the norm and a blockage of the identical. This, to my mind, is what was at the heart of the economic crises of the beginning of the twentieth century—*crises of normalization*, of the emplacement of repetition—when mass production began to demand a radical recasting of the industrial apparatus. For recording is indeed inscribed as the death of representation.

Right from the beginning, machines invented to counteract temporal erosion, to constitute a *speech* that would be indefinitely *reproducible*, to overcome the ravages of time by means of the construction of mechanical devices, were moving in the direction of a death blow to representation. Let us listen to the first androids built in the eighteenth century by the Abbot Mical:

> The King brings peace to Europe
> Peace crowns the King with glory

> And peace is the contentment of the people
> Oh cherished King, father of your people, let
> Europe behold the glory of your reign.[101]

Beginning with the first mechanically produced discourse, the repetitive concretization of the unalterable has always taken the initial form of a renewed affirmation of power and legitimate might.

A revealing situation: recording and the reproduction of speech reconstitute the locus of power. Through the androids, it is authority itself that is speaking. Authority, and simultaneously, *paradoxically*, its *caricatured* double. For in droning the discourse of the established powers, these androids simulate them, mimic them. This raises a scandalous question: are those powers not also copies, simulacra that are themselves susceptible to simulation?

Thus the simulation of the master's word leads to a questioning of the status of the master himself. Mechanisms for recording and reproduction on the one hand provide a technical body, a framework for representations, and on the other hand, by presenting themselves as a *double*, constitute a simulacrum of power, destroy the legitimacy of representation.

The first recording of speech was a representation of the king's legitimacy. But the android of the king, repeating his legitimacy, could not remain a representation of power for long.

Chapter Four
Repeating

The power to record sound was one of three essential powers of the gods in ancient societies, along with that of making war and causing famine. According to a Gaelic myth, it was precisely by opposing these three powers that King Leevellyn won legitimacy.[102]

Recording has always been a means of social control, a stake in politics, regardless of the available technologies. Power is no longer content to enact its legitimacy; it records and reproduces the societies it rules. Stockpiling memory, retaining history or time, distributing speech, and manipulating information has always been an attribute of civil and priestly power, beginning with the Tables of the Law. But before the industrial age, this attribute did not occupy center stage: Moses stuttered and it was Aaron who spoke. But there was already no mistaking: the reality of power belonged to he who was able to reproduce the divine word, not to he who gave it voice it on a daily basis. Possessing the means of recording allows one to monitor noises, to maintain them, and to control their repetition within a determined code. In the final analysis, it allows one to impose one's own noise and to silence others: "Without the loudspeaker, we would never have conquered Germany," wrote Hitler in 1938 in the *Manual of German Radio*.

When Western technology, at the end of the nineteenth century, made possible the recording of sound, it was first conceived as a political auxiliary to representation. But as it happened, and contrary to the wishes of its inventors, it invested music instead of aiding institutions' power to perpetuate themselves; everything suddenly changed. A new society emerged, that of mass production, repetition, the nonproject. Usage was no longer the enjoyment of present labors, but the consumption of replications.

Music became an industry, and *its consumption ceased to be collective*. The hit parade, show business, the star system invade our daily lives and completely transform the status of musicians. Music announces the entry of the sign into the general economy and the conditions for the shattering of representation.

This major conflict, inherent in industrial society, between the logic of industrial growth and the political exigencies of the channelization of violence, was announced in the confrontation between the repetitive penitents of Lent and the differentiated masks of Carnival. The fundamental answer: to *silence*, through a monologue of organizations distributing normalized speech.

For with the appearance of the phonograph record, the relation between music and money starts to be flaunted, it ceases to be ambiguous and shameful. More than ever, music becomes a monologue. It becomes a material object of exchange and profit, without having to go through the long and complex detour of the score and performance anymore. Capitalism has a frank and abstract interest in it; it no longer hides behind the mask of the music publisher or entertainment entrepreneur. Once again, music shows the way: undoubtedly the first system of sign production, it ceases to be a mirror, an enactment, a direct link, the memory of past sacrificial violence, becoming a solitary listening, the stockpiling of sociality.

The mode of power implied by repetition, unlike that of representation, eludes precise localization; it becomes diluted, masked, anonymous, while at the same time exacerbating the fiction of the spectacle as a mode of government. *Music announces that we are verging on no longer being a society of the spectacle*. The political spectacle is merely the last vestige of representation, preserved and put forward by repetition in order to avoid disturbing or dispiriting us unduly. In reality, power is no longer incarnated in men. It is. Period.

The emergence of recording and stockpiling revolutionizes both music and power; it overturns all economic relations.

By the middle of the twentieth century, representation, which created music as an autonomous art, independent of its religious and political usage, was no longer sufficient either to meet the demands of the new solvent consumers of the middle classes or to fulfill the economic requirements of accumulation: *in order to accumulate profit, it becomes necessary to sell stockpileable sign production, not simply its spectacle*. This mutation would profoundly transform every individual's relation to music.

Just as the street hawker's blue books shaped the reader and supplanted the storyteller, just as the printer supplanted the copyist, representation would be replaced by repetition, even if for a time it looked as though they had reached an accommodation. Like the others, this shake-up was ineluctable. Once music became an object of exchange and consumption, it hit against a limit to accumulation that only recording would make it possible to exceed. But at the same time, repetition reduces the commodity consumption of music to a simulacrum of its original, ritualistic function, even more so than representation. Thus *the growth of exchange is accompanied by the almost total disappearance of the initial usage of the exchanged.* Reproduction, in a certain sense, is the death of the original, the triumph of the copy, and the forgetting of the represented foundation: in mass production, the mold has almost no importance or value in itself; it is no longer anything more than one of the factors in production, one of the aspects of its usage, and is very largely determined by the production technology.

Reproduction, then, emerges as a tremendous advance, each day giving more people access to works created for representation—formerly reserved for those who financed the composition of the work—than at any other time since man's creation. But it also entails the individualization of the sacrificial relation as a substitute for the simulacrum of the rituality of music.

This constitutes, moreover, a massive deviation from the initial idea of the men who invented recording; they intended it as surface for the preservation of representation, in other words, a protector of the preceding mode of organization. It in fact emerged as a technology imposing a new social system, completing the deritualization of music and heralding a new network, a new economy, and a new politics—in music as in other social relations.

In the eighteenth century, the paradigm of representation succeeded in establishing itself as a scientific method in music and the sciences. Economic theories, political institutions, and counterpowers were born of these theories: the practice of creating economic models, combinatorics, harmony, the labor theory of value and the theory of social classes, Marxism. All of these concepts stem from the world of representation and still live by its conflicts. Recording expresses itself in an overturning of the whole of understanding. Science would no longer be the study of conflicts between representations, but rather the analysis of processes of repetition. After music, the biological sciences were the first to tackle this problem; the study of the conditions of the replication of life has led to a new scientific paradigm which, as we will see, goes to the essence of the problems surrounding Western technology's transition from representation to repetition. Biology replaces mechanics.

For the turn of the century was the moment when programs for the repetition of man and his discourse became generalized, shattering speech and differences, in order to channel violence and the imaginary into commodity needs and false subversions.

This radical mutation was long in the making and took even longer to admit. Because our societies have the illusion that they change quickly, because the past slips away forgotten, because identity is intolerable, we still refuse to accept this most plausible hypothesis: if our societies seem unpredictable, if the future is difficult to discern, it is perhaps quite simply because *nothing happens, except for the artificially created pseudoevents and chance violence that accompany the emplacement of repetitive society.*

In this type of organization of the production of society, power can no longer be located simply in the control of capital or force. It is no longer an enactment through representation. And if there are no longer any localizable power holders, neither are there counterpowers that can be institutionalized in response. Power is incorporated into the very process of the selection of repeatable molds. It is spread among the different elements of the system. Impossible either to locate or seize, having become the genetic code of society, power must be changed or destroyed.

Music, transformed into a commodity, gives us insight into the obstacles that were to be encountered by the ongoing commodification of other social relations. Music, one of the first artistic endeavors truly to become a stockpileable consumer product, is exemplary. However, we must avoid reading this as a global plot of money against sociality. Neither money nor the State entirely understood or organized this mutation of music and its recording. The first, beginning in the nineteenth century, they saw only as a harmless diversion, and the second only as a functional tool to make the leader's work easier. The history of the process of the emplacement and generalization of recording is thus the history of an invention which, in spite of its inventors, played a far-reaching role in the restructuring of society. Conceived as a way of preserving one network (representation), recording was to create another (repetition), and heralded an immense mutation in knowledge and politics.

The Emplacement of Recording

Freezing Speech

In a half-century's time, an invention that was meant to stabilize a mode of social organization became the principal factor in its transformation. Beyond music, the process of this technological and ideological mutation brought on an entire transformation of a paradigm and a world vision. I would like to describe the conditions of this birth in enough detail to make its scale apparent, to allow us to ponder the real conditions of insertion of an invention in a mode of social organization, conditions that are very often unrelated to those anticipated by the innovators themselves.

In the second half of the nineteenth century, when industry established the economy of its reign, at least two French recording procedures (Léon Scott's phonotaugraph and Charles Cros' paleophone) preceded Edison's cylinder-based phonograph, which would ultimately gain acceptance. Both of them failed because they did not demonstrate the economic advantages of their use in repre-

sentation, or perhaps because they did give a glimpse of the economic advantages of the rupture of that network and of the profitability of replication.
Their projects, like Edison's later on, had the same aim as printing; they were
designed to transform sound into writing, in other words, to achieve automatic
stenography. Scott's phonotaugraph, developed toward 1861, transcribed
sounds onto a disk covered with lampblack; a typographer, he went unheard.
Cros' invention, developed around the same time, had no more success. After
registering it with the Academy of Sciences on April 30, 1877, he wrote:

> There is every reason to believe that they wanted to sidetrack me and
> I had the foresight to have my sealed envelope opened. . . . Justice
> will be done in the long run, perhaps, but in the meantime these things
> remain an example of the scientific tyranny of the capital. They ex-
> press this tyranny by saying: theories float in the air and have no
> value, show us some experiments, some facts. And the money to run
> the experiments? And the money to go look at the facts? Get what you
> can. It is thus that many things are not carried out in France.[103]

Charles Cros, who died in poverty on August 9, 1888, was faulted for not
being a specialist: the author of *Hareng saur* (*Kippered Herring*) and the creator
of the *Groupe des hydropathes* (Hydropaths' Group) could not be taken seriously scientifically. But more profoundly, these inventors' failure is undoubtedly tied to the fact that no one sensed that the society of the time had a
need to communicate more extensively than it was already doing. Representation
sufficed to regulate the imaginary in the channelization of violence. Thus when
the first transatlantic telegraph line was laid between London and New York,
Emerson remarked, "But will we have anything to say to each other?" There
was as yet no solvent market for the recording of representation.

For the same reasons, Edison's phonograph, patented on December 19,
1877, was not significantly developed either, until people began to realize that
there could be a discourse saleable on a society-wide basis. Edison himself, who
after 1878 lost interest in it and turned to promoting the electric light bulb, presented his invention as a stenographic machine for the reproduction of speech,
for recording discourse, the purpose of which was to stabilize representation
rather than to multiply it. The emphasis was thus placed on *preservation*, not
mass *replication*. The first phonographs functioned as recorders used on a very
localized basis to preserve and transmit exemplary messages. Edison considered
that this usage in itself was enough to justify the economic exploitation of his
invention: "We will be able to preserve and hear again, one year or one century
later, a memorable speech, a worthy tribune, a famous singer, etc. . . . We
could use it in a more private manner: to preserve religiously the last words of
a dying man, the voice of one who has died, of a distant parent, a lover, a mis-
tress." Complaints such as these, "Oh, if only we had Mirabeau's speeches, or

Danton's, etc.,'' would become impossible. In conformity with this prognosis, the phonograph was first used to disseminate the voices of leaders (Kossuth, Gladstone, etc.), in other words, as an archival apparatus for exemplary words, a channelization of the discourse of power, a recording of representation, of the boss's orders. It also showed up in the preacher's pulpit and the teacher's office. In fact, speech was the only sound it was technically feasible to record before 1910, and even then only a few operas were recorded. It was not until 1914 that the first symphony was recorded (Beethoven's Fifth, directed by Artur Nikish).

The phonograph was thus conceived as a privileged vector for the dominant speech, as a tool reinforcing representative power and the entirety of its logic. No one foresaw the mass production of music: the dominant system only desired to preserve a recording of its representation of power, to preserve itself. For this reason, there were various attempts during this period to use recording to constitute a language of the international elite that would "transcend" national differences and make it possible to give world status to the preservation of representation, thus creating a real, solvent market for the recordings.

The attempts to transcribe music into language or language into music reflect this will to construct a universal language operating on the same scale as the exchanges made necessary by colonial expansion: music, a flexible code, was dreamed of as an instrument of world unification, the language of all the mighty. For example, in one of the most talked-about essays of the nineteenth century, François Sudre, a French engineer, presented a procedure for the formation of a musical language. The Académie des Beaux-Arts de l'Institut, in its report of 1827, found that "the author perfectly fulfilled the goal he set out to accomplish. Providing men with a new means of communicating their ideas to one another, of transmitting them long distances and in the deepest night, is a true service to society." In Sudre's musical language, the seven notes of the scale could be used to express any idea.[104] Using only three notes, Sudre devised telephony, in other words, "the art of using the sounds of an instrument to send from a distance signals transmitting orders, dispatches, and phrases inscribed in advance in a special vocabulary . . . designed to conform to the range of the regulation bugle and to adapt it to military art." The idea of a language coded in music is linked to the idea of military order and imperial universality. Similarly, on February 21, 1891, in the Grand Amphitheater of the Ecole des Hautes Etudes Commerciales (the location is symptomatic) *Volapuk*, or "World Language" (from *vol* for "world" and *pük* for "speak") was unveiled: exchange still thought itself capable of imposing a universal language as a space for the production of messages recorded and distributed worldwide, and of making the phonograph a privileged auxiliary of this strategy of the existing powers.

In an article entitled "L'industrie phonographique aux Etats-Unis" ("The Phonograph Industry in the United States"), we find a fairly accurate analysis of the way in which the use of the phonograph for the recording and distribution of music was excluded, or allowed a very limited role, in the conceptions of the nineteenth-century inventors:

To prevent the quest for large profits from compromising future profits, an article published in *The Phonogram* proposes that each local company create two separate divisions, one to occupy itself with the trinkets, the other to take care of the serious side. It informs us that there exists in New Jersey a veritable music factory issuing several rolls of new tunes each month. These tunes, moreover, are of many varieties, depending on the nature of the customers. Before the piece is recorded, the title is shouted into the machine. After it is performed, if a little space is left on the cylinder, they make a point of using it to record the applause and cheers that the musicians lavish upon themselves at the end. All of the pieces are played before being put on the market; those exhibiting defects are set aside. The pieces sell for one to two dollars each, and some profit is left after the musicians' salaries are subtracted, because the same piece is inscribed on several cylinders at the same time. The earnings of the phonographs used in the "nickel in the slot" system vary greatly according to the location of the machine and the nature of the piece, which is changed every day. Certain machines have yielded up to fourteen dollars in daily earnings.[105]

But the promoters themselves considered this usage of recording to be secondary, and Edison opposed using the phonograph for it, particularly in the form of jukeboxes in drugstores, because he thought it might make it "appear as though it were nothing more than a toy." In 1890, he wrote:

In my article of twelve years ago I enumerated among the uses to which the phonograph would be applied: 1. Letter-writing and all kinds of dictation, without the aid of a stenographer. 2. Phonographic books, which would speak to the blind people without effort on their part. 3. The teaching of elocution. 4. Reproduction of music. 5. The "Family Record," a registry of sayings, reminiscences, etc., by members of a family, in their own voices: and of the last words of dying persons. 6. Music boxes and toys. 7. Clocks that should announce, in articulate speech, the time for going home, going to meals, etc. 8. The preservation of languages, by exact reproduction of the manner of pronouncing. 9. Educational purposes: such as preserving the explanations made by a teacher, so that the pupil can refer to them at any moment; and spelling or other lessons placed upon the phonograph for convenience in committing to memory. 10. Connection with the telephone, so as to make that invention an auxiliary in the transmission of permanent and invaluable records, instead of being the recipient of momentary and fleeting communications. Every one of these uses the perfected phonograph is now ready to carry out. I may add that, through the facility with which it stores up and reproduces music of all sorts, or whistling and recitations, it can be employed to furnish constant amusements to invalids, or to social assemblies, at receptions, dinners, etc. . . . Music by a band—in fact, whole operas—can be stored up on the cylinders, and the voice of Patti singing in England can thus be heard again on this side of the ocean, or preserved for future generations.[106]

An incredible text: the inventor himself criticizing what was to become the major use to which his invention would be put, namely the reproducibility, the accessibility, the sociality of music. It was not until 1898 that he realized the commercial potential for recorded music.

Edison was not the only one who was off the track. Even musicians saw this tool only as a secondary technique allowing for a slight improvement in the conditions of representation.

In 1903, the artists and entertainment professionals questioned in France in a survey on the gramophone declared that they were happy that, thanks to it, their ephemeral interpretations would survive, and they would thus be the equals of those who inscribe their work in a permanent medium, such as composers and writers. Theater directors saw it as a way of eliminating noise from the wings and of pressuring musicians in the case of a strike. Orchestra leaders expected it to be no more than a pedagogical aid. For example, M. Luigini, the conductor at the Opera, declared on March 24, 1902: "I deem that this instrument is called upon to play an important role as *educator*, through the dissemination of the works of the masters as performed by the best interpreters. It will be an invaluable instruction, popularizing the good traditions of purity of style of the performing elite." But later on, when people began to understand the uprooting it would cause, the conservatives were fraught with worry. The phonograph was then seen as something dangerous, giving a wide audience effortless access to a consumption of signs reserved for an elite: "With the phonograph, as with the automatic piano or organ," declared the president of the Commission pour la Rénovation et le Développement des Etudes Musicales (Commission for the Renovation and Development of Musical Studies) in 1930, "one may derive profound pleasure with no study whatsoever." This brings to mind a slip of the pen in the *Monitor*, which, in transcribing a speech by Villèle on the "democratization" of art, spoke of its "demoralization."[107]

If at first everything seemed to be going against the use of recording for the distribution of music, that would gradually change. And at first with an apparently minor invention improving upon the technology. But an underground innovation is sometimes more decisive than the tool it perfects: just as the telephone could not have survived without the introduction of the commutator, electroforming, used by Tainter in 1886 to perfect the graphophone, would prove far more central to the emergence of repetition than the phonograph itself, because it made serial repetition possible (the cylinder wore out after six playings). Conceived as an instrument functioning for power, the recorder did not definitively enter the apparatus of consumption until after electroforming signed its apparent depoliticization, and exemplary words were replaced by repetition and accessibility.

That would take time. In 1887, the American Graphaphone Company was founded; it commercialized the first cylinders for use in amusement parks and as dictating machines in government agencies. Berliner invented the flat 78-RPM record in 1902, and the double-sided record in 1907. The first concert broadcast over the radio was on June 15, 1920, at Chelmsford. The first commercial success for a record was in 1925 ("Let It Rain, Let It Pour"), when the introduction of electrical recording considerably improved the technology. It was distributed through the jukebox, which created demand prior to the existence of a private market for record players; the jukebox in fact constituted a collective consumption, a final form of the concert guaranteeing the transition between representation and the solitary consumption organized by the record player, which did not develop until after the Great Depression, the Second World War, and the invention, in 1948, of the long-playing microgroove Vinylite record.

The phonograph, then, is part of a radically new social and cultural space demolishing the earlier economic constructions of representation. With the introduction of the record, the classical space of discourse collapses. Against the wishes of Edison himself, the drugstore jukebox wins out over the singers of the *caf' conç*, the record industry over the publishing industry. Even radio, which could have forestalled this process by providing representation with a new market, gradually became, as we will see, an auxiliary of the record industry. After the discourse of representation was devalued, radio provided a showcase for the record industry, and the record industry gave radio the material it needed to fill the airwaves.

The American Graphophone Company, which was more willing than its competitors to admit this rerouting in the usage of the invention, prevailed in the beginning by emphasizing the disk over the cylinder, reproduction over recording, music over speech.

From the beginning, it was necessary to produce demand at the same time as producing the supply. Thus the Compagnie Française du Gramophone (French Gramophone Company) starting in 1907, organized representations of repetitions, free musical shows in all the towns. A journalist was amazed by "the possibility this opens of being able to listen to a repertory composed of works from all periods, by the best performers in the entire world." The gramophone seemed powerful and original because, since it *plugged into a stockpile playing on time and space*, it seemed to be a tool for the generalization of representation, a symbol of the internationalization of social relations.

During this period, several innovations were made that completed the rerouting of the original invention. After the electrical recording of orchestra music began, there developed a competition between rotation speeds. While the conflict between the disk and the cylinder concluded with the victory of one of the techniques over the other, this time it ended in compromise: the 78-RPM record disappeared, but two speeds persisted where, technologically, one would have been enough. This throws light on capitalism's new face: it is no longer enough for an innovation to be objectively better for it to be marketed and replace the others. It is possible to retard its introduction, eliminate it, or only partially adopt it, if necessary. The reason is simple: the record object is not usable by itself. Its use-value depends on that of another commodity, because repetition requires a reproduction device (the phonograph). A major modification of the repeated object would be enough to make the reproduction device obsolete, and vice versa; prudence in innovation becomes necessary, and the economic process loses its fluidity.

This interdependence of use-values would gradually become generalized, in an economy in which practically every repetition requires a duality of the used object and the "user": film and camera, light bulb and lamp, blade and razor, automobile and highway, detergent and washing machine. This duality is, as we will see, characteristic of the economy of repetition and is responsible for retarding its evolution.

Records and Radio

Until 1925, the record was very little used; the waxes were of bad quality and transmission was only possible by placing the microtransmitter close to the phonograph's acoustical horn, resulting in very bad transmission. In 1925, these two disadvantages were overcome by electrical recording, the use of better waxes, and the invention of the pickup, permitting direct transmission from the record. In the beginning, there were no problems associated with using records: record producers freely distributed their products to the various radio stations. But two or three years later, complaints against the use of records for radio broadcasting began to be made; they were first voiced by music writers, music publishers, performing artists, and above all record manufacturers.

The *authors* said nothing at first, thinking that radio broadcasts gave them good publicity. But later, they began to fear that the public would lose interest in performance halls and that record sales would fall as public broadcast cut into private consumption. The *music publishers* saw one of their markets shrinking. They were the ones who sold the scores to radio musicians, a declining market. Moreover, music publishers had an interest in record sales; at that time, they held reproduction rights that they exercised against record manufacturers. The *performing artists* saw one of their places of work disappearing. The *record manufacturers* also feared a decline in record sales. They held rights to the mechanical reproduction of works under copyright, which they acquired either from the author or the publisher. In the second case, which was the most common, their reproduction rights were strictly limited to reproduction by mechanical means; they were not given general reproduction rights, which remained in the hands of the publisher. Thus it was impossible for the record manufacturers to oppose radio broadcast, which, as already noted, was considered a form of representation: they could neither invoke copyright law nor claim unfair competition to prevent their product from being used. We will see later on that this problem has yet to be resolved even today.

Exchange-Object and Use-Object

Reproduction did not have a dramatic impact on the economic status of music until sixty years after its introduction. The existing copyright laws, which defined a musical work as something written and attributed an exchange-value to its representation, provided no answer to these questions: Can a phonograph cylinder be considered a "publication" protected as such under law? In other words, does sound recording entail a right for the person whose work is recorded? What share of the exchange-value of the recording should go to the creator? Will he continue to be a rentier, as he was under representation? What compensation should be given for playing a record, in other words, for the representation of repetition, the use of the recording? How should the performers and companies who made the recording be compensated?

These questions were not unique to music: at least as early as the beginning of the nineteenth century, the problem of ownership rights over the reproduction of a work was posed for all of the arts, and more generally for all productions for which there exists a technology permitting the replication of an original. These questions became central when competitive capitalism and the economy of representation catapulted into mass production and repetition.

In France, the law of July 19, 1793, did not make it clear whether income from reproduction should go to the creator of the original or to the person who buys his work. However, this issue took on considerable economic importance in the course of the nineteenth century; it was decided on a case-by-case basis, depending on the balance of power between creators and merchants. Painting is a case in point: Watteau died in poverty while the engravers of his works made fortunes; Léopold Robert, on the other hand, in one year sold a million prints of *Les Moissonneurs* (*The Reapers*), the original of which was bought for 8,000 francs; Gérard made 40,000 francs from his *Bataille d'Austerlitz* (*Battle of Austerlitz*), which it cost Pourtalès 50,000 francs to have engraved; Ingres, an able manager of his financial interests and glory, ceded the reproduction rights for his works for 24,000 francs.[108]

In France, a law of May 16, 1886, regulated the specific case of "mechanical instruments," or recording without prior representation or a preexisting score—barrel organs, music boxes, pianolas, aeolian harps—which had made considerable headway as substitutes for bands at dances. The rights belonged to the industrialist and not the musician: "The manufacture and sale of instruments serving mechanically to reproduce copyrighted musical tunes do not constitute musical plagiarism as envisioned under the law of July 19, 1793, in combination with articles 425ff. of the Penal Code." Royalties could be collected only when these machines were used for public representations.

The manufacturers of these machines unsuccessfully cited this text to support their refusal to pay royalties to the authors of songs they reproduced, insisting on the private nature of the use to which these instruments were put. A decision of November 15, 1900, returned a guilty verdict against a café owner who installed a pay music box in his establishment.[109] But can this same reasoning be applied to records, equating them with scores?

In this period, the compensation performers and authors obtained, in the absence of texts, varied widely and was in some cases considerable. In 1910, Mme. Melba and Tamagno received 250,000 and 150,000 francs respectively from the Compagnie des Gramophones; Caruso made a fortune on his recordings beginning in 1903.

The courts had a hard time settling on an interpretation of the law. The civil court of the Seine district, in a ruling of March 6, 1903, authorized the recording of music without payment of any royalties. Then a decisive ruling was issued on February 1, 1905, by the court in Paris: its motivation is very interesting, since it was one of the last attempts to maintain the fiction of representation in repetition, of the written in the sonorous, in order to equate the record with the score, which requires a specialized knowledge to be read:

Finding that disks or cylinders are impressed by a stylus under which they pass; that they receive a graphic notation of spoken words, that the thought of the author is as though materialized in numerous grooves, then reproduced in thousands of copies of each disk or cylinder and distributed on the outside with a special writing, which in the future will undoubtedly be legible to the eyes and is today within everyone's reach as sound; that by virtue of this repetition of imprinted words, the literary work penetrates the mind of the listener as it would by means of sight from a book, or by means of touch with the Braille method; that it is therefore a mode of performance perfected by performance, and that the rules of plagiarism are applicable to it.[110]

An astonishing text: it equates the record with the score. Written reproduction determines the record's exchange-value and justifies the application of copyright legislation. It should also be noted that in this judgment sound reproduction is considered a popular by-product of writing, anticipating a time when specialists would decipher the recording directly.

The problem of compensating authors and performers was passed over in silence by legislators for a long time: in his fascinating report of 1910 to the

Chamber of Deputies, Th. Reinach does not once refer to the case of music. Then, little by little, the principle of copyright was established for records. Authors and certain performers became the recipients of rent, a result of laws or court judgments on the mechanical reproduction of their works. The parallel to writing was pursued, and institutions were established to regulate this industrial production on behalf of those receiving rent. These associations, similar to author's associations, enforced payment of royalties to the authors, performers, and publishers of the works. In effect, repetition poses the same problems as representation: rent presupposes a right to industrial production, in other words, a right to monitor the number of pressings and number of copies sold, to which the royalties are proportional. And it also presupposes the confidence of the authors, who totally delegate the management of their economic rights to experts working for associations whose function it is to valorize their works.

The author's associations thus played a decisive role in determing the relations between music and radio. In law, the radio broadcast of a work was deemed a public performance on July 30, 1927 (by decision of the criminal court of Marseilles),[111] and consequently the law of 1791 became applicable. There was a fleeting attempt to develop another position. L. Bollecker, in a 1935 article in the *Revue Internationale de Radioélectricité*, makes a distinction between radio broadcasting, which consists in transmitting waves through space, and reception, which consists in transforming those waves into sounds. In this view, only reception is representation, and it is generally private (it would only be public if the loudspeaker were public). Radio broadcasting, for its part, would be a new form of publishing. An extraordinary fantasy of spatial writing, the marking of space. Bollecker, however, was not followed: since waves are not *durable* and are *immaterial*, radio broadcasting remained a form of representation.

The opposition to the use of records on radio was resolved in France by a contract between SACEM and the private stations concluded in 1937. After that, radio stations had to pay for representation and reproduction rights.

The performers and publishers would continue to be excluded. They were recognized as having no claim. The difficulties associated with the evaluation of copyrights and related claims in representation resurface here, because the multiplicity of sale and listening sites make it difficult to collect payment.

In addition, there arise specific obstacles to monitoring recordings, because free access is taken to a new height: today it has become possible for each listener to record a radio-broadcast representation on his own, and to manufacture in this way, using his own labor, a repeatable recording, the use-value of which is a priori equivalent to that of the commodity-object, without, however, having its exchange-value. This is an extremely dangerous process for the music industry and for the authors, since it provides free access to the recording and its repetition. Therefore it is fundamental for them to prevent this diversion of usage, to reinsert this consumer labor into the laws of commercial exchange, to suppress information in order to create an artificial scarcity of music. The simplest solution would be to make such production impossible by scrambling the quality of the broadcast representation, or by truncating it, or again by taxing this independent production, financing royalty payments on these unknown recordings through a tax on tape recorders—this is done in Germany. The price of music usage is then based entirely on the price of the recorder. But the number of recordings could increase without a change in the number of tape recorders. We could then conceive of a tax on recording tape, which would mean paying music royalties in proportion to the exchange-value of nonmusic.

This problem of monitoring recording already announces a rupture in the laws of the classical economy. The independent manufacture of recording, in other words, consumer labor, makes it more difficult to individualize royalties and to define a price and associated rent for each work. It is conceivable that, at the end of the evolution currently under way, locating the labor of recording will have become so difficult, owing to the multiplicity of the forms it can take, that authors' compensation will no longer be possible except at a fixed rate, on a statistical and anonymous basis independent of the success of the work itself.

At the same time, usage becomes transformed, *accessibility replaces the festival*. A tremendous mutation. A work that the author perhaps did not hear more than once in his lifetime (as was the case with Beethoven's Ninth Symphony and the majority of Mozart's works) becomes accessible to a multitude of people, and becomes repeatable outside the spectacle of its performance. It gains availability. It loses its festive and religious character as a simulacrum of sacrifice. It ceases to be a unique, exceptional event, heard once by a minority. The sacrificial relation becomes individualized, and people buy the individualized use of order, the personalized simulacrum of sacrifice.

Repetition creates an object, which lasts beyond its usage. The technology of repetition has made available to all the use of an essential symbol, of a privileged relation to power. It has created a consumable object answering point by point to the lacks induced by industrial society: because it remains at bottom the only element of sociality, that is to say of ritual order, in a world in which exteriority, anonymity, and solitude have taken hold, music, regardless of type, is a sign of power, social status, and order, a sign of one's relation to others. It channels the imaginary and violence away from a world that too often represses language, away from a representation of the social heirarchy.

Music has thus become a strategic consumption, an essential mode of sociality for all those who feel themselves powerless before the monologue of the great institutions. It is also, therefore, an extremely effective exploration of the past, at a time when the present no longer answers to everyone's needs. And above all, it is the object that has the widest market and is the simplest to promote: *after the invention of the radio, that incredible showcase for sound objects, sol-*

vent demand could not but come their way. It was inevitable that music would be instituted as a consumer good in a society of the sonorous monologue of institutions.

The use-value of the repeated object is thus the expression of lacks and manipulations in the political economy of the sign. Its exchange-value, approximately equal today for every work and every performer, has become disengaged from use-value. Ultimately, the price bears no direct relation to the production price of the record itself, to the quality, properly speaking, of the recording. It depends very heavily on the process of the production of the demand for music and on its fiscal status, in other words, on the role assigned to it by the State.

Exchange-Time and Use-Time

Repetition constitutes an extraordinary mutation of the relation to human production. It is a fundamental change in the relation between man and history, because it makes the stockpiling of time possible. We have seen that the first repetition of all was that of the instrument of exchange in the form of money. A precondition for representation, money contains exchange-time, summarizes and abstracts it: it transforms the concrete, lived time of negotiation and compromise into a supposedly stable sign of equivalence in order to establish and make people believe in the stability of the links between things and in the indisputable harmony of relations.

Repetition goes much further, when reproduction becomes possible for an object and no longer only for the standard: with the stockpiling of music, a radically new economic process got under way. It was thought that discourse—in other words, exchange-time once again—was being stockpiled, while in fact what was being stockpiled was coded noise with a specific ritual function, or use-time. For we must not forget that music remains a very unique commodity; to take on meaning, it requires an incompressible lapse of time, that of its own duration. Thus the gramophone, conceived as a recorder to stockpile time, became instead its principal user. Conceived as a word preserver, it became a sound diffuser. The major contradiction of repetition is in evidence here: *people must devote their time to producing the means to buy recordings of other people's time*, losing in the process not only the use of own their time, but also the time required to use other people's time. Stockpiling then becomes a substitute, not a preliminary condition, for use. *People buy more records than they can listen to. They stockpile what they want to find the time to hear.* Use-time and exchange-time destroy one another. This explains the valorization of very short works, the only ones it is possible to use, and of complete sets, the only ones worth the effort of stockpiling. This also explains the partial return to a status prior to that of representation: music is no longer heard in silence. It is integrated into a whole. But as background noise to a way of life music can no longer endow with meaning.

Double Repetition

The network of repetition is indissociable from the nature of the musical code that is transmitted within it. The music conveyed within repetition, except for the valorization of the stockpile bequeathed by representation and music that continues to be created in the representative network, is in fact repetitive music in the literal sense.

Today, music emerges above all in its commodity component, in other words, as popular song, commercialized by radio. The remainder of production, learned music, is still inscribed within the theoretical line of representation and its crisis; it constitutes, in appearance, a totally different field from which the commodity is excluded and in which money is not a concern. But in reality it is not that at all: the rupture of the code of harmony leads to an abstract music, noise without meaning. In contrast to previous centuries, popular music and learned music, the music of above and the music of below, have broken their ties with one another, just as science has broken its ties with the aspirations of men. Nevertheless, their subterranean connections remain very deep. They are both in effect products of the rupture of the system of representation, and one of the most interesting problems in the political economy of music is interpreting the simultaneity of a *fracture* in meaning, and of the emplacement of repetition or the *absence* of meaning. A fracture shattering all of political economy and heralding the emplacement of repetition, its lacks, and its coming crisis.

Mass Repetition: The Absence of Meaning

The mass-produced music that surrounds us is the product of an industry. Since the first commercial records and their success following the Great Depression, and the invention, supported by radio broadcasting, of the long-playing record in the 1940s, the increasing mechanization of musical production has dramatically changed the conditions and meaning of that music. Looking at it from the outside, we have the impression of witnessing the birth of an ordinary consumer industry in which the process of capitalist concentration should be functioning. But it is much more than that, a very different situation.

Of course, capital is more of a presence than ever. Producing a record requires considerable funds and sophisticated technology, and the potential for profits has led several financial groups to develop an interest in music. For example, Gulf and Western, the oil, real estate, and cigarette conglomerate, bought out Volt (with the singer, Otis Redding), and the Transamerica Insurance Company took over Liberty and World Pacific (with Ravi Shankar). Still, the reality cannot be reduced to a passage from competitive capitalism (which would be representation) to monopoly capitalism (which would be repetition). The economy of music, a strange industry on the borderline between the most sophisticated marketing and the most unpredictable of cottage industries, is much more original and much more of an augur of the future than that.

First, use-value in the music industry does not depend on the product alone, but also on the use-value of the receiver available to consumers. It is therefore impossible to make a rapid change in broadcast technology. Thus the competition between producers cannot be based simply on the quality of the product, or even on price, because the products are too diversified to be comparable.

Secondly, the production, strictly speaking, of the object (the record) is only a minor part of the industry, because the industry, at the same time as creating the object of exchange, must also create the conditions for its purchase. It is thus essentially an industry of manipulation and promotion, and repetition entails the development of service activities whose function is to produce the consumer: the essential aspect of the new political economy that this kind of consumption announces is the *production of demand*, not the *production of supply*. Of course, it is hard to admit that the value of the object is not in the work itself, but in the larger whole within which the demand for commodities is constructed. Nevertheless, we will see that this has been the new logic of the economy of music from the moment it constituted itself as an industry, directly after the Second World War.

Producing the Market: From Jazz to Rock

Music did not really become a commodity until a broad market for popular music was created. Such a market did not exist when Edison invented the phonograph; it was produced by the colonization of black music by the American industrial apparatus. The history of this commodity expansion is exemplary. A music of revolt transformed into a repetitive commodity. An explosion of youth —a hint of economic crisis in the middle of the great postwar economic boom— rapidly domesticated into consumption. From Jazz to Rock. Continuations of the same effort, always resumed and renewed, to alienate a liberatory will in order to produce a market, that is, supply and demand at the same time.

In the slang of the black ghettos, ''to jazz'' and ''to rock'' both meant to make love. Significantly, they were lived, festive acts; they became neuter commodities, cultural spectacles for solvent consumers.

Jazz was strategically situated to produce this market: it had never been able to constitute a commercial object under representation, because, as an unwritten music almost entirely tied to very localized cultures and audiences, it lacked a solvent market. It came to expression in the United States, where the largest market of solvent young people would be born and where recording technology would be produced. A music of the body, played and composed by all, jazz expressed the alienation of blacks. Whites would steal from them this creativity born of labor and the elementary forms of industrialization, and then turn around and sell it back: the first market for jazz was composed of the uprooted black workers of the ghettos of the Northern cities. White capital, which owned all of the record companies, controlled this commercialization process from the start, economically and culturally.

Significantly, the first jazz record was recorded by a white band (the Original Dixieland Jazz Band). The economic appropriation of jazz by whites resulted in the imposition of a very Westernized kind of jazz, molded by white music critics and presented as music "accessible to the Western musical ear"—in other words, cut off from black jazz, allowing it to reach the white youth market:

> Jazz is cynically the orchestra of brutes with nonopposable thumbs and still prehensile toes, in the forest of Voodoo. It is entirely excess, and for that reason more than monotone: the monkey is left to his own devices, without morals, without discipline, thrown back to all the groves of instinct, showing his meat still more obscene. These slaves must be subjugated, or there will be no more master. Their reign is shameful. The shame is ugliness and its triumph.[112]

In the early days of jazz recording, moreover, the best-known jazz musicians were white: [Paul] Whiteman (elected "King of Jazz" in 1930), Benny Goodman (the sacred "King of Swing"), Stan Kenton. Starting in the 1930s, when the demand for blues became heavy enough to incite hopes of a profit, production was systematically developed through the prospecting and pillaging of the patrimony of the southern blacks: the idea of paying royalties to blacks did not occur very often to those who recorded their songs. The system of "field trips" (collecting tours through the South organized by procurers—sometimes black—called talent scouts) made it possible to furnish the newly arrived migrants in the big northern industrial cities a standardized reflection of the musical forms of their culture of origin.

> The traveling studios recorded as much material as possible, gave each singer a few dollars, and that was that. Did the records sell well or not? The artists didn't know anything about it and could never have profited from it. Only the stars were called up North to record regularly; but there again, they were paid by the piece, not by the sales.[113]

This exploitation of black musicians would continue for a long time and is still going on. Many blacks profited from this system and accepted this way of making their music known. As Adorno writes, jazz "took pleasure in its own alienation," frankly reflecting the situation of blacks, accepting their exploitation until the end of the 1950s. After 1955, the commodification of jazz was confirmed; the rhythm and blues of the black ghettos of the North reached white circles, thanks to the massive introduction of 45s and specialized AM radio station programming.

The 78-RPM record disappeared and 45s took over, thanks to the jukebox.

An enormous, unified, standardized market was developed, centered on high school styles.

In addition, the baby boom and the end of the postwar economic crisis produced an enormous demand on the part of white youth, coincident with the introduction of a syncretic product ready to respond to that demand by using black despair—carefully filtered—to express young white hopes: *rock*. A very precise filtering carried out simultaneously by the radio stations and ASCAP (American Society of Composers, Authors, and Publishers), which monitors royalty payments.

The style established itself, exploding in the 1960s, and mass production began. Another stage was then reached with the entry of genuine black artists, paradoxically reimported to America by English groups or expatriate American blacks in England (such as Jimi Hendrix). This solidified with the development of the 33-RPM record and the FM radio station network, which in the United States have very largely replaced 45 records and AM stations.

Just as the Romantic explosion called the harmonic code into question, the explosion of the 1960s nearly called into question the standardized pop music/rock market. But where representation had been shattered, repetition would prevail: silencing people is possible in repetition, but not in representation. The record industry controlled what the late nineteenth-century music publishers were unable to prohibit. Explicit censorship played a very prominent role, readopted from the eighteenth century. The Jefferson Airplane, for example, were fined $1,000 on several occasions for not honoring clauses in their contracts prohibiting verbal abuse; the Grateful Dead were fined $5,000 in Texas for the same reason; Country Joe McDonald was fined $500 and sentenced to having his head shaved for uttering the word ''fuck'' in Massachusetts; Jim Morrison got six months' imprisonment and a fine of $500 for ''indecent exposure'' and ''use of offensive language'' in Florida.[114]

Thus a degraded, censored, artificial music took center stage. Mass music for an anesthetized market.

The Production of Supply

In repetition, the entire production process of music is very different from that of representation, in which the musician remained the relative master of what he proposed for the listener. He alone decided what to do. Of course, as soon as sound technology started to play an important role in representation, the musician was already no longer alone. But today, under repetition, the sound engineer determines the quality of the recording, and a large number of technicians construct and fashion the product delivered to the public. Thus the decision to go back to a recording to perfect it or leave it as is is the prerogative of at least two tiers of sound technicians, whose criteria for the decision are obviously very different from those of the performers and authors. The performer is only one element contributing to the overall quality; what counts is the clinical purity of the acoustics. The result is a profound mutation of the aesthetic criteria in relation to those of representation. The record listener, conditioned by these production criteria, also begins to require a more abstract form of aesthetics. Sitting in front of his set, he behaves like a sound engineer, a judge of sounds.

Little by little, the very nature of music changes: the unforeseen and the risks of representation disappear in repetition. The new aesthetic of performance excludes error, hesitation, noise. It freezes the work out of festival and the spectacle; it reconstructs it formally, manipulates it, makes it abstract perfection. This vision gradually leads people to forget that music was once background noise and a form of life, hesitation and stammering. Representation communicated an energy. Repetition produces information free of noise. The production of repetition requires a new kind of performer, a virtuoso of the short phrase capable of infinitely redoing takes that are perfectible with sound effects. Today, the examples of this sometimes extreme dichotomy between great performers who put their names on records they did not record, and great technicians who, on stage, are incapable of moving the audience that buys their records, are becoming increasingly common.

What is more, the dichotomy is taken to the point of schizophrenia: Janis Joplin's backup bands and the "Chaussettes noires" ("Black Socks") were not composed of the same musicians on stage and in the studio. Elisabeth Schwartz-kopf agreed to record in Kirsten Flagstad's name.[115] The author disappears behind the interpreter, who starts calling the shots, demanding to be cosignatory and to have exclusive rights. Musicians and performers separate into two classes: the stars of repetition—disembodied, ground up, manipulated, and reassembled on record—and the anonymous functionaries of the local festival, minor bands, residue of representation. Of course, there are transition points: the popular dance is the school for stars, the place where they find an audience. But the two are linked to radically different modes of production and radically different statuses: a festival status and a penitence status, Carnival and Lent.

The Production of Demand: The Hit Parade

Outside of a ritual context or a spectacle, the music object has no value in itself. It does not acquire one in the process that creates supply, because mass production erases value-creating differences; its logic is egalitarian, spreading anonymity and thus negating meaning. Value may then base itself, partially or totally, on an artificial, unidimensional differentiation, the only thing left allowing hierarchy, ranking. That is why the hit parade is so important to the organization of the commercialization of music. Hit parades, a subtle mystification, play a central role in this new type of political economy. Far from being a superfluous reflection of the gadget economy, a publicity stunt, or a neutral market indicator, they are, to my mind, the prime movers of the repetitive economy, a herald of new processes on the way, the end of the market economy and the price system.

In the case of semiidentical objects selling for the same price and arriving on the market in quantities so high that radio programmers, let alone the consumers, cannot test them all, differentiation requires a ranking scheme exterior to their production, one recognized as legitimate by the consumers and capable of defining for them the use-value of the title. For the use-value of a song is not only reflected, but also created by its place on the hit parade: a title that no longer ranks has no use-value. It is therefore essential that the consumer believe in the legitimacy of this hierarchy, which reflects and creates value. Thus the hit parade must appear to be both an expression of sales figures (this was the case for the first hit parades of the 1930s, undertaken very seriously by *Cash Box* and *Billboard*; even today, access to the information is restricted to manufacturers) and a prediction of future success. The result is an ambiguous mixture based on sales figures, with listener preference supposedly playing a role in the ranking.

It is thus a mode of evaluation of exchange-value, of relative price, that relates both to the market (sales) and planning (the elective process). *The hit parade bears a relation to the dream-form of the socialist economy, in which price is no longer the sole determining factor of use/exchange value, in which choice is expressed not only by disposable income but by the democratically expressed preferences of the consumers.*

But this dream of a way out of the commodity system and the rules of capitalism is, quite the opposite, their most modern and accomplished form. For the ranking is never more than mythically a reflection of the desires of the consumers. It is not based solely on the work's satisfaction of an audience's mysterious and elusive tastes. Those who believe they participate in the ranking by writing to radio stations and newspapers should know that in many cases their letters are not even opened, or their telephone calls made note of. The ranking, in fact, depends very largely, on the one hand, on pressures applied on station program planners by record manufacturers eager to see their new releases carve out a place for themselves on the market and, on the other hand, on real sales results. The speed of a title's climb up the charts is thus very largely a function of the number and quality of the new titles awaiting release. It has "value" in the eyes of the listeners, then, by virtue of the ranking to which they think they contributed. When, as usually happens, they buy in quantities proportional to the rankings, they justify them, bringing the process full circle.

Thus the hit parade system advertises the fact that the value of an object depends on the existence of other, alternate objects, and disappears when the possibility of making more surplus-value with other objects arises. An object's value is a function of the intensity of the financial pressures of the new titles waiting to enter circulation. Exchange thus completes the destruction of any fiction of autonomous, stable use-value: *usage is no longer anything more than the public display of the velocity of exchange*. I do not mean to say that the hit parade creates the sales, but, much more subtly, that it channels, selects, and gives a value to things that would otherwise have none, that would float undifferentiated.

It is clear that the hit parade is not unique to music. It can serve to establish a hierarchy of use-values for any and all of the identically priced objects that are flooding our lives in increasing numbers. An economy in which these products assail the consumer, so that only a small number of them have a chance of retaining his attention, necessitates a public display of relative values, something the pricing system can no longer signify.

Public display requires a display board, in other words, a media system capable of periodically publicizing the rankings, in order to organize a rapid turnover of the objects. Once again, the radio-broadcast network is at the heart of the manipulation. Radio is necessary to the record industry's success, just as the record industry conditions the profitability of radio. More exactly, the success of a sound network depends on its capacity to sell music objects; it appears increasingly to be the case that a radio station has an audience only if it broadcasts records that sell. Stations do not create record sales, but they do reassure those who, in any case, have decided to buy them. In addition, on the more general level, only what speaks to purchasers about purchasing gets heard or read these days. Radio has become the showcase, the publicity flier of the record industry, like women's magazines for women's consumption and photo magazines for camera consumption—when it does not have, in addition, a financial stake in the success of the works by way of usurped royalties.

Of course, radio is also judged by its capacity to inform; but this is not irrelevant to our topic. For a secondary function of the hit parade is to create a pseudoevent, in a repetitive world in which nothing happens anymore. We can even go so far as to say that since the emergence of the hit parade, all that radio broadcasts any more is information: *on the spectacle of politics in newscasts, on objects in advertisements, and on music in the hit parades.*

Let there be no mistake: I do not believe that industry fabrication of a "hit" is possible today, or that a record ranked first on the hit parade and played several times a day will necessarily sell better than records that are not. There are innumerable counterexamples of records that are "plugged" unsuccessfully, or of records forsaken by program planners that sell in the hundreds of thousands. But now that the number of new releases is approaching one hundred a day, the success selection process is statistical, fortuitous. The best will find their way. Production is thus still determined by the craftsman and his group, tied in with the industrial apparatus. Selection and turnover are aided and channeled by the media network and the hit parades. An evolution in the publishing system[116] has led to the disappearance of the primary function of the publisher, namely to undertake the general promotion of the work;[117] instead, that function is entrusted to specialized distributors or even to the radio stations, and the door is left open for all manner of pressure tactics and all the ruses of corruption. In an economy in which the production price of the supply is very low in relation to the production price of the demand, continuity of expansion largely depends upon the improvement of commercialization techniques. The marketing of music is very different from that of the other products of individual consumption. Leaving aside mass-produced music, for which night clubs offer an effective market, low sales restricted to very special, localized markets can easily be made profitable without extensive market research. Moreover, the process of the production of demand accelerates the production of supply: the management of the stockpiles becomes increasingly the duty of specialists. The risks of selection are eliminated by specialized intermediaries (''rock jobbers''), who organize the rapid rotation of the stockpiles and the concentration of sales on a few big successes outside the specialized networks, thus accelerating the movement of songs on the hit parades.

The Banalization of the Message

It is not that song has become debased; rather, the presence of debased songs in our environment has increased. Popular music and rock have been recuperated, colonized, sanitized. If the jazz of the 1960s was the refuge of a violence without a political outlet, it was followed by an implacable ideological and technical recuperation: Jimi Hendrix was replaced by Steve Howe, Eric Clapton by Keith Emerson. Today, universalizing, despecifying degradation is one of the conditions for the success of repetition. The most rudimentary, flattest, most meaningless themes pass for successes if they are linked to a mundane preoccupation of the consumer or if they signify the spectacle of a personal involvement on the part of the singer. The rhythms, of exceptional banality, are often not all that different from military rhythms. To judge by its themes, neither musically nor semantically does pop music announce a world of change. On the contrary, nothing happens in it anymore, and for twenty years it has seen only very marginal, or even cyclic, movement. Change occurs through the minor modification of a precedent. Each series is thus repeated, with slight modifications enabling it to parade as an innovation, to constitute an event. The singers of the 1950s are back in fashion in the 1970s, and today's children enjoy their parents' records. At times, however, the quality improves, song becomes critical and music blasphemous: *repetitive, detached*, as though denouncing standardization; it heralds a new subversion by musicians cramped by censorship, who stand alone in announcing change.

The Confinement of Youth

Even though it concerns all social categories and every specialized market, mass music in the new form of the repetitive economy is first and foremost a process for channelization of childhood. Little by little, it establishes the youth as a separate, adulated society with its own interests and its own culture different from that of adults, its own heroes and battles. In fact, even in the idealized form of the Beatles' docile pseudorevolt, it assured that young people were very effectively socialized, in a world of pettiness constructed by adults. The cultural universe of this music produced by adults organizes group uniformity. The music is experienced as relation, not as spectacle; as a factor of unanimity and exclusion in relation to the world of adults, not as individual differentiation.

The life of their dreams is a "pop life," a refuge from the great uncontrollable machines, a confirmation of the individual's sameness and the collectivity's powerlessness to change the world. The music of repetition becomes both a relation and a way of filling the absence of meaning in the world. It creates a system of apolitical, nonconflictual, idealized values. It is here that the child learns his trade as a consumer, for the selection and purchase of music are his principal activities. One might say that this is the child's labor, his contribution to production: an upside-down world in which the child produces the consumption of music while industry produces the demand for it. The production of supply becomes secondary and, with it, musicians.

Music thus fashions a consumer fascinated by his identification with others, by the image of success and happiness. The stars are always the idealized age of their audience, an age that gets younger as the field of repetition expands. This channelization of childhood through music is a politically essential substitute for violence, which no longer finds ritual enactment. The youth see it as the expression of their revolts, the mouthpiece of their dreams and needs, when it is in fact a channelization of the imaginary, a pedagogy of the general confinement of social relations in the commodity.

Undoubtedly, musical repetition—an essential element in consumer initiation, the consumption of discourse with a simplistic code—will some day make necessary a definitive rupture with the mode of socialization it replaces. Therefore *mass music heralds the death of the family*: when it seems to serve the interests of commodity society to make the child a direct employee of the State, or to send him out into the job market at the age of twelve to guarantee him a minimal salary and to make him solvent, repetitive society will doubtless be skillful enough to make claims of progress for this new linkup.

Repetition offers another challenge to the analysis of the behavior of agents in classical economics and Marxism: musical consumption leads to a sameness of the individual consumers. One consumes in order to resemble and no longer, as in representation, to distinguish oneself. *What counts now is the difference of the group as a whole from what it was the day before, and no longer differences within the group.* This socialization through identity of consumption, this mass production of consumers, this refusal of what in the recent past was a proof of existence, goes far beyond music. Thus we are seeing women's fashion disappear in the uniformity of jeans; we are seeing new generations buy records and clothes (in other words, their sociality) in huge anonymous retail outlets where mass production is shamelessly displayed, where children come, fascinated by the Pied Piper of Hamelin.

Background Noise

Mass music is thus a powerful factor in consumer integration, interclass leveling, cultural homogenization. It becomes a factor in centralization, cultural normalization, and the disappearance of distinctive cultures.

Beyond that, it is a means of silencing, a concrete example of commodities speaking in place of people, of the monologue of institutions. A certain usage of the transistor radio silences those who know how to sing; the record bought and/or listened to anesthetizes a part of the body; people stockpile the spectacle of abstract and too often ridiculous minstrels.

But silencing requires the general infiltration of this music, in addition to its purchase. Therefore, it has replaced natural background noise, invaded and even annulled the noise of machinery. It slips into the growing spaces of activity void of meaning and relations, into the organization of our everyday life: in all of the world's hotels, all of the elevators, all of the factories and offices, all of the airplanes, all of the cars, everywhere, it signifies the presence of a power that needs no flag or symbol: musical repetition confirms the presence of repetitive consumption, of the flow of noises as ersatz sociality.

This situation is not new. After all, Haydn and Mozart's works were almost exclusively background music for an elite who valued them only as a symbol of power. But here power has extended its functions to all of society and music has become background noise for the masses. The music of channelization toward consumption. The music of worldwide repetition. Music for silencing.

This makes repetitive society easier to analyze than the society that preceded it: in imposing silence through music, it speaks of itself. The organizations betray their strategies. Not much effort is required to hear the repressive role of mass music.

Take, for example, one of the most characteristic firms dealing in the music of silencing: Muzak. Created in 1922 to provide music over the telephone, it branched out beginning in 1940 into selling atmosphere music. It has countless clients: stadiums, parks, salons, cemeteries, factories, clinics (including veterinary clinics), banks, swimming pools, restaurants, hotel lobbies, and even garbage dumps.

The pieces of music used on the tapes they sell are the object of a treatment akin to castration, called ''range of intensity limitation,'' which consists of dulling the tones and volume. They are then put on perforated cards, classed by genre, length and type of ensemble, and programmed by a computer into sequences of 13 1/2 minutes, which are in turn integrated into completed series of eight hours, before being put on the market.

According to David O'Neill, one of Muzak's executives: "We do not sell music; we sell programming." For a restaurant, "the breakfast programs ordinarily consist of recent tunes without too much brass. For lunch, we mainly put on songs with string accompaniment." For a factory or office: "The current should go against the flow of professional fatigue. When the employee arrives in the morning, he is generally in a good mood, and the music will be calm. Toward ten thirty, he begins to feel a little tired, tense, so we give him a lift with the appropriate music. Toward the middle of the afternoon, he is probably feeling tired again: we wake him up again with a rhythmic tune, often faster than the morning's."

This music is not innocent. It is not just a way of drowning out the tedious noises of the workplace. It may be the herald of the general silence of men before the spectacle of commodities, men who will no longer speak except to conduct standardized commentary on them. It may herald the end of the isolable musical work, which will have been only a brief footnote in human history. This would mean the extermination of usage by exchange, the radical jamming of codes by the economic machine. This is given explicit approval by musicians who think music should insinuate itself into the everyday world and cease to be an exceptional event. John Cage, for example, writes:

> Nevertheless, we must bring about a music which is like furniture—a music, that is, which will be part of the noises of the environment, will take them into consideration. I think of it as melodious, softening the noises of the knives and forks, not dominating them, not imposing itself. It would fill up those heavy silences that sometimes fall between friends dining together. It would spare them the trouble of paying attention to their own banal remarks. And at the same time it would neutralize the street noises which so indiscreetly enter into the play of conversation. To make such music would be to respond to a need.[118]

Is Cage simply speaking of atmosphere music, or does he see this in the long run as a strategy for the radical destruction of usage in music, a politics of the liquidation of meaning, opening the way for a subsequent renaissance? So-called learned music, which is the context of his remarks, arrives at the negation of meaning announced by mass music.

Repetition and the Destruction of Meaning

The present state of music theory in the West is tied, through its discourse, to the ideological reorganization necessitated by the emplacement of repetition. The theoretical musician acts within this logic, however ambiguous it may be. Like the musician in representation, he remains a musician of power, paid to perfect the sound form of today's technical knowledge, while at the same time signifying its liquidation. there is no contradiction in that: *the absence of meaning is the necessary condition for the legitimacy of a technocracy's power.*

The musician, once outside the rules of harmony, tries to understand and master the laws of acoustics in order to make them the mode of production of a new sound matter. Liberated from the constraints of the old codes, his discourse becomes nonlocalizable. Pulverizer of the past, he displays all of the characteristics of the technocracy managing the great machines of the repetitive economy. He is under the regime of nonsense and shares all of its attributes:

Scientism. Western music theory is expressed essentially in the context of its relation to science and its crisis: "Our period will be occupied, and for several generations, with the construction and structuration of a new language, which will be the vehicle of the masterpieces of the future" (Boulez). Or again: "Music is unified with the sciences in thought. Thus, there is no break between the sciences and the arts. . . . Henceforth, a musician should be a manufacturer of philosophical theses and global systems of architecture, of combinations of structures (forms) and different kinds of sound matter" (Xenakis).

The parallel to science is total. Like science, music has broken out of its codes. Since the abandonment of tonality, there has been no criterion for truth or common reference for those who compose and those who hear. Explicitly wishing to create a style at the same time as the individual work, music today is led to elaborate the criterion of truth at the same time as the discovery, the language (*langue*) at the same time as speech (*parole*). Like science, music then moves within an increasingly abstract field that is less and less accessible to empiricism, where meaning disappears in abstraction, where the dizzying absence of rules is permanent. Thus music voices the becoming of science in repetition, and its difficulties. It is linked to an abstraction of the conditions of functioning of the society taking root, of the difficulties of a science of repetition.

Imperial universality. An elite, bureaucratic music—for the moment still without a commercial market, supported by powers in search of a language, of a project—it desires to be universal, as they are. In order to be universal, it diminishes its specificity, reduces the syntax of its codes. It does not create meaning: for *meaninglessness is the only possible meaning in repetition without a project.* Music even tries to be the general theory of all structures. Take Xenakis: "Musicians could have, for the benefit of nineteenth-century physics, created the abstract structure of the kinetic theory of gases, exclusively by and for music." "Musical thought lagged far behind thinking in physics and mathematics, an avant-garde cut off from philosophy, thus chastised. We have decided that is necessary for it to catch up with them to lead them once more, as at the time of its Pythagorean birth." Or Stockhausen: "Without question, *I want to integrate everything.*" This is the power vocabulary of the managers of industrial society, of men of learning persuaded of the existence of an all-encompassing truth, of a society that *desires to make its simple management the matrix* of its meaning. A frenzied search for universal abstraction by men whose labor has lost its meaning and who are incapable of finding a more exalting one for it than the statistical organization of repetition.

Depersonalization. The music of power no longer conveys information within a code. It is, like the ideology of the period, without meaning. The modern musician says nothing, signifies nothing if not the insignificance of his age, the impossibility of communication in repetition. He no longer claims to communicate with the listener by means of a message, although this void sometimes leaves room for a message, because the listener can make associations or try to create his own order in the void. "We facilitate the process so that anything can happen" (Cage). Musical production is no longer configured. Meaningless, music is the source of silence, but also of creative emergence; it rejects the hypothesis of a natural foundation for relations of sound, refuses "a natural organization of sensible experience."[119] Form is freed from the constraint of having a single configuration and is founded upon an infinite labyrinth of "feedback" effects.[120]

The deconcentration and manipulation of power. The noise of matter, unformed, unsaleable, confirms the negation of meaning. This ideology of nonsense is not without political ramifications. In fact, it heralds the ideology of repetitive society, the simulacrum of the decentralization of power, a caricature of self-management. All of music becomes organized around the simulacrum of nonpower. Instead of the score and orchestra leader, improvisation is presented as the form of composition. In fact, the most formal order, the most precise and rigorous directing, are masked behind a system evocative of autonomy and chance. The performer himself is only falsely associated with the elaboration of the music. Philip Glass, for example, writes music allowing the musicians to take an active role. The music is performed by a permanent group, in which the instrumentalists are seated facing one another in a circle. There is no orchestra leader; Glass simply gives a nod of his head to start or pass from one part to another. However, despite these appearances, the musician has never been so deprived of initiative, so anonymous. The only freedom left is that of the synthesizer: to combine preestablished programs. A simulacrum of self-management, this form of interpretation is a foreshadowing of a new manipulation by power: since the work lacks meaning, the interpreter has no autonomy whatsoever in his actions; there is no operation of his that does not originate in the composer's manipulation of chance. Managing chance, drawing lots, doing anything at all, consigns the interpreter to a powerlessness, a transparency never before achieved: he is an executor bound by laws of probability, like the administrator in a repetitive society. His status is thus not innocent but, once again, premonitory. The place of an individual in the modern economy is no different from that of Glass' interpreter: whatever he does, he is no more than an aleatory element in a statistical law. Even if in appearance everything is a possibility for him, on the average his behavior obeys specifiable, abstract, ineluctable functional laws. Behind the disorder of the theory, then, lies a music of the mean, of anonymity reconstituted within a context of general individuality, a music inducing us to hear "the voice as one of the things of the universe" (Michel Serres), background noise for a repetitive and perfectly mastered anonymity.

As Adorno writes: "The irrationality of its impact, calculated to the extreme, seeks to keep people on a leash, a parody of protest against the supremacy of the classificatory concept." The mutation of the combinatory field, the opening of great sound spaces, and the control exerted over the performers do not, however, give the composer unlimited power. Instead of toying with the limited nomenclature of the harmonic grid, he outlines processes of composition, experiments with the arrangement of free sounds. An acoustician, a cybernetician, he is transcended by his own tools. This constitutes a radical inversion of the innovator and the machine: instruments no longer serve to produce the desired sound forms, conceived in thought before written down, but to monitor unexpected forms. Bach recreated the organ to fit his music; the modern composer, on the other hand, is now rarely anything more than a spectator of the music created by his computer. He is subjected to its failings, the supervisor of an uncontrolled development.

Music escapes from musicians. Even if they believe that they conceive structures and theorize on the basis of their experience (or rather, on a tangent to it)—using often erroneous borrowings from misunderstood mathematical theory —their role is only to guide the unpredictable unfolding of sound production, not to combine foreseeable sounds originating from stable instruments. "The composer becomes a kind of pilot pushing buttons."[121] He becomes the organizer ("program designer") of a fluid work: quasi-alive, quasi-organic, creating its own signification through its history. A work no one except the composer can perform, often for lack of instruments specially invented for it. Music, like political economy—and once again far in advance of it—experiences the transcending of men by their knowledge and tools: one no longer organizes the growth of production, but instead only makes an effort to regulate an evolution whose laws are not completely known (that of sound matter). One produces what technology makes possible, instead of creating the technology for what one wishes to produce.

Elitism. This depersonalization in statistical scientism results in the elimination of style and at the same time the demand for its impossible recovery, the search for an inimitable specificity ("Anything with an easily imitated style appears suspect to me").[122] It then becomes a question of writing music in uncultured forms, thus running the risk of floating off into generalized statistical abstraction. In repetition, the group that engineers the repetition only preserves itself as an autonomous group by artificially distancing itself from the rest of humanity, by being alone in understanding such intellectualized music, by continually revising its own system for deciphering signs.

These musicians demonstrate the frailty of this strategy of the power elite: the strategy of continually attempting to invent a new code precludes any chance of producing one that could remain stable. Accepting nonsense as the foundation of its power, it precludes any chance of its developing an acceptable rationale for its domination.

The absence of styles necessitates permanent innovation in titles (Nono's *Incontri*, Cage's *Construction*, Kagel's *Match*, Foss' *Time Cycle*) and competition outside the traditional forms of musical production. The music exists, imposes itself, without seeking to meet listeners' demand. This is indeed the pure ideology of progress: value in itself, even if it destroys the use-value of communication.

Scientism, imperial universality, depersonalization, manipulation, elitism—all of the features, all of the foundations of a new ideology of the political economy are already present in contemporary Western musical research. A music without a market imposes itself on an international elite, which once again finds itself exceeding national cultural traditions, seeking the Esperanto it needs to function smoothly, to communicate effectively: the dream of achieving a worldwide unity of the great organizations through the language of music, a language that finds its legitimacy in science and imposes itself through technology. Recording and preservation raised hopes for this universality in 1859 (with the international agreement on the definition of *la*) and 1880 (with Volapuk), and atonal abstraction is in the process of bringing it to realization: a transnational and transideological convergence around a shared loss of mastery over production to a scientifically based abstraction. Around shared power founded on a principle of nonideology.

Musicians have become the symbol of this nonideological multinationalism: esteemed in all of the most cosmopolitan places of power, financed by the institutions of the East and West, they are the image of an art and science common to all of the great monologuing organizations. Even though the modern musician, because he is more abstract, gives the appearance of being more independent of power and money than his predecessors, he is, quite the opposite, more tightly tied in with the institutions of power than ever before. Separated from the struggles of our age, confined within the great production centers, fascinated by the search for an artistic usage of the management tools of the great organizations (computer, electronic, cybernetic), he has become the learned minstrel of the multinational apparatus. Hardly profitable economically, he is the producer of a symbolism of power.

Theoretical music liquidates; it confirms the end of music and of its role as a creator of sociality. Therefore power, which is supported by it, confirms, utilizes, and is founded upon the end of meaning.

A music for which the visual and relational aspect is without question very secondary, an abstract music, it accommodates itself very well to the technology of the network of repetition. There is no reason to believe that the elite will not someday even succeed in imposing its consumption, pure nonsense, upon the subjects of the repetitive world. It will perhaps be the next generation's substitute for pop music. In repetition, then, music is no longer anything more than a slightly clumsy excuse for the existence of musicians, who are theorists and ideologues—just as the economy becomes the excuse, gradually forgotten, for the power of the technocrats.

The Linking of the Two Circulations: The Culmination

The linkup of these two productions may seem a priori artificial: one uses the most traditional of harmonies to avoid startling anyone, while the other is inscribed in an abstract search, in a theoretical corpus in crisis, and refuses to accept the dominant trends and cultural codes. One addresses itself to a mass audience with the aim of inciting it to buy, the other has no market or financial base other than patronage, public or private. Yet both belong to the same reality, that of hyperindustrialized Western society in crisis. They are thus necessarily linked to one another, if only by virtue of being radical opposites.

In fact, their interlinkage is much more solid than this simple antithesis implies: theoretical music and mass music both relate to a repetitive image of Western society, in which the necessity of mass production legitimates the technocracy that universally engineers, manipulates, and distributes it. They reveal that standardization and technocracy are two aspects of universalized repetitiveness, and that science loses its meaning at the same time as the commodity sheds its usage. This culmination may be the precondition for the birth of a new music beyond the existing codes; it may also be the herald of a new dictatorship of representation and the emplacement of a new dominant code, a universal functionalism. It is preparing the ground for the destruction of all past symbolic systems, all earlier networks for the channelization of the imaginary—to allow the network of a dictator to impose itself or, on the contrary, so that each individual may create his own network.

The Concerts of Power

The interdependence of the political functions of these very different forms of music is everywhere to be heard. For example, the concert hall, an invention of the eighteenth century, is still an instrument of power today, regardless of what kind of music is offered—just as the museum remains a political substitute for merchants in the management of art.

The concert, the central site of representative society, remains operative in repetitive society. But the spectacle is more and more in the hall itself, in the audience's power relation with the work and the performer, not in its communion with them: today, a concert audience judges more than it enjoys; music has become a pretext for asserting one's cultivation, instead of a way of living it. This is necessary continuance in repetitive society of the enactment of the political, in which the spectators recognize their own image and the legitimation of their own power on stage and in the hall.

How many errors would have been avoided in social science over the past two centuries if it had known how to analyze the relations between spectators and musicians and the social composition of the concert halls. A precise reflection of the spectators' relation to power would have been seen immediately. It would have been seen that representation in the concert hall changes dramatically with the emplacement of repetition; that the elite defines and protects itself through esotericism and the cultural level required for the works it listens to. The concert is then seen to be the place used by the elite to convince itself that it is not as cold, inhuman, and conservative as it is accused of being. For the rest, the concert is mediocrity camouflaged in an artificial festival, for "whatever is distributed to the poor can never be anything more than poverty."[123] Concerts of popular music, tours by artists, are now all too often nothing than copies of the records, the cold perfection of which they try to recreate through the generalized practice of lip syncing. The popular dance, which has in part become a concert, is a release for violence that has lost its meaning. Carnival without the masks and the channeling of the tragic; in which the music is only a pretext for the noncommunication, the solitude, and the silence imposed by the sound volume and the dancing; in which even in its worldly substitute, the night club, the music prevents people from speaking—people who in any event do not want to, or cannot, speak. For them, there is already silence in repetition.

The Musician of Power: From Actor to Molder

The musician has become an element in a new network of power. Except in cases in which he is identified with a single work, the musician is much better known than the music he writes or performs. In a general way, the performer has eclipsed the author and often even steals his creation. In this way he crystallizes the last forms of the spectacle, which are necessary to make repetitive society tolerable. His function is no longer to invent ways of communicating or representing the world, but to be a *model for replication*, the mold within which reproduction and repetition take shape. This explains the importance of justifying his function through science, through the style he creates, or through the idealization of his personal image. This explains the importance of the process of identification: beyond time, etched in the object, the artist becomes the replicated mode. His function is no longer musical, but unifying. He is one of the genes necessary for repetition. When the spectacle dissolves in replication, the author-performer becomes a mold. The Beatles and David Bowie have played and continue to play this role. Whether they are manipulated, exploited by intermediaries who manufacture them, or are masters of their own game, they are still models whose life style and clothing are replicated. Thus they continue to play the eternal role of music: creating a form of sociality. But in repetition that passes for identity, and no longer for difference: *the scapegoat has become a model.*

Language itself is a mold: the conversation of the consumers is stereotyped, restricted to the words if not the titles of the songs, and becomes an element in this radical identification. One might wonder whether atonal music could eventually play the same role. Presently, its only connection with repetition is that it is statistical, macromusical. But after all, it serves as a model for the elite group, and it may one day expand, becoming a general attribute of the spectacle. And, as we have already in some way indicated, atonal music, conceived for abstract listening, is infinitely better suited to recording and replication than harmony, conceived for the spectacle.

Replication is thus, in a subtle way, at the origin of a strange festival, where all masks are identical. Of a Carnival among the penitents of Lent. Of a Carnival that is tragic because mimicry is fully at play. The process of replication functions even in cases in which the intent is critical or the identification is made with a nonconformist model; anticonformism creates a norm for replication, and in repetition music is no longer anything more than a detour on the road to ideological normalization.

The Delocalization of Power

The modes of the accommodation of music to power undergo dramatic change in the course of this process. Whereas representation constituted a complex hierarchy of styles and musical consumption, the two kinds of music we find in repetition have one and the same function, that of general leveling, the power elite excepted. In representation, music was endowed with a social prestige in the eyes of the middle classes that was all the greater because it incarnated the cultural values of the upper classes while allowing the middle classes to distinguish themselves from the poor—for example with the piano, which for the bourgeoisie was a means of gaining access to a simulacrum of the representation of music and to romantic culture. This disappears in repetition.

First, popular music is no longer hierarchically organized according to class. It is the same at the top and the bottom of the social scale, because the media have considerably reduced the time it takes for a success to penetrate socially and geographically, as well as reducing its life span. In the popular dances and night clubs of the world's capitals, it is increasingly the same music that is heard, and same dances danced. But it is no longer the case, as it was in the Middle Ages, that inspiration flows from the people to the courts; instead, the markets that the industrial apparatus addresses are becoming uniform.

Secondly, learned music is restricted to a concert-going elite. It is no longer distributable among the middle bourgeoisie, since the instruments and techniques it uses make it impossible for amateurs to communicate or perform it. It is a kind of music that is limited to specialists in the aleatory, a spectacle organized by technicians for technocrats.

The game of differentiation through consumption, creator of individualist desire, still continues in relation to "hi-fi systems." But I think that once the period of technological fascination is over, a banalization of these objects and an identification with the group in mass repetition will set in, as it did with automobiles and records.

Thus music has become an element in the normalized reproduction of the labor force and of social regulation. By that token, it is simultaneously Order and Transgression, a support for Lent and a Carnival substitute. As the technology of musical distribution changed, the deritualization of music accelerated, its artificial role nearly disappeared, and, beyond its spectacle, a music of identity took root.

Taking the analogy further, we might say that the listener in front of his record player is now only the solitary spectator of a sacrificial vestige. Doubtless, a hereditary memory of the process preserves music's power of community, even when it is heard in solitude. But the disappearance of the ceremony, and even the sacrificial spectacle, destroys the entire logic of the process: there is no longer a closed arena of sacrifice, the ritual or the concert hall. The threat of murder is everywhere present. Like power, it slips into homes, threatening each individual wherever he may be. Music, violence, power are no longer localized in institutions.

When this happens, music can no longer affirm that society is possible. It repeats the memory of another society—even while culminating its liquidation—a society in which it had meaning. In the disappearance of the channeling sacrifice and the emergence of repetition, it heralds the threat of the return of the essential violence. Thus, from whichever direction we approach it, music in our societies is tied to the threat of death.

Repetition, Silence and the End of Sacrifice

Repetition and Silence

It is not the least of the paradoxes of our research that we have detected uniformity in such multiform music, repetition in a society that talks so much about change, silence in the midst of so much noise, death in the heart of life. Everywhere, in fact, diversity, noise, and life are no longer anything more than masks covering a mortal reality: Carnival is fading into Lent and silence is setting in everywhere.

First, it must be stated that mass production compels silence. A programmed, anonymous, depersonalized workplace, it imposes a silence, a domination of men by organization. The people who manufacture mass-produced commodities have neither the means nor the time to speak to one another or experience what they produce. In the time of representation, of individualized production in competitive capitalism, the work existed and took form in a concrete, lived time. Today, neither the musician nor the worker who produces the record on an automatic machine have the time to experience the music. Mass production is programming, the monotonous and repeated noise of machines imposing silence on the workers. This silence in replication becomes more pronounced with the growing automation of the process. Fewer and fewer people work at machines for the silent manipulation of entrapped speech. An extraordinary spectacle: the double silence of men and commodities in the factory. An intense spectacle, because after leaving the factory the commodities will speak much more than the people who manufactured them.

Silence is the rule in consumption as well: the mass repetition and distribution of uniform models, interchangeable with money, totally obstructs communication by way of object-related differences. Thus we are almost outside of the society of representation, in which people communicated to one another through the different objects they used. Identity-spending, the difference-creating Carnival mask—both of which are desired in hierarchical societies—are in the process of *being replaced by successive waves of collective nondifferentiation.* Unanimity becomes the criterion for beauty, just as in the hit parade the criterion of usage is confused with the quantity sold. Power, which in representation is delegated, in repetition is appropriated by a knowledge-wielding minority.

Contrary to currently fashionable notions, the triumph of capitalism, whether private or State, is not that it was able to trap the desire to be different in the commodity, but rather that it went far beyond that, making people accept identity in mass production as a collective refuge from powerlessness and isolation. Capitalism has become "a terrorism tempered by well-being, the well-being of each in his place" (Censor).[124] For with records, as with all mass production, security takes precedence over freedom; one knows nothing will happen because the entire future is already laid out in advance. Identity then creates a mimicry of desires and thus rivalry; and once again repetition encounters death. Today's return to violence is therefore not caused by an excessive will to difference, but on the contrary by the mass production of mimetic rivalries and the absence of anything serving to focus this violence toward a sublimating activity.

The emplacement of general replication transforms the conditions of political control. It is no longer a question of making people believe, as it was in representation. Rather, it is a question of Silencing—through direct, channeled control, through imposed silence instead of persuasion.

This strategy is not new: I showed earlier that royal power allowed the written press to develop when it came to see the press as a way of channeling rumors and of replacing libels and tracts. Today, repetitive distribution plays the same role for noise that the press played for discourse. It has become a means of isolating, of preventing direct, localized, anecdotal, nonrepeatable communication, and of organizing the monologue of the great organizations. One must then no longer look for the political role of music in what it conveys, in its melodies or discourses, but in its very existence. Power, in its invading, deafening presence, can be calm: people no longer talk to one another. They speak neither of themselves nor of power. They hear the noises of the commodities into which their imaginary is collectively channeled, where their dreams of sociality and need for transcendence dwell. The musical ideal then almost becomes an ideal of health: quality, purity, the elimination of noises; silencing drives, deodorizing the body, emptying it of its needs, and reducing it to silence. Make no mistake: if all of society agrees to address itself so loudly through this music, it is because it has nothing more to say, because it no longer has a meaningful discourse to hold, because even the spectacle is now only one form of repetition among others, and perhaps an obsolete one. In this sense, music is meaningless, liquidating, the prelude to a cold social silence in which man will reach his culmination in repetition. Unless it is the herald of the birth of a relation never yet seen.

Noise Control

The absence of meaning, as we have said, is nonsense; but it is also the possibility of any and all meanings. If an excess of life is death, then noise is life, and the destruction of the old codes in the commodity is perhaps the necessary condition for real creativity. No longer having to say anything in a specific language is a necessary condition for slavery, but also of the emergence of cultural subversion.

Today, the repetitive machine has produced silence, the centralized political control of speech, and, more generally, noise. Everywhere, power reduces the noise made by others and adds sound prevention to its arsenal. Listening becomes an essential means of surveillance and social control.

Today, every noise evokes an image of subversion. It is repressed, monitored. Thus, the prohibition against noise in apartment buildings after a certain hour leads to the surveillance of young people, to a denunciation of the political nature of the commotion they cause. It is possible to judge the strength of political power by its legislation on noise and the effectiveness of its control over it. In addition, the history of noise control and its channelization says much about the political order that is being established today.

Before the Industrial Revolution, there existed no legislation for the suppression of noise and commotion. The right to make noise was a natural right, an affirmation of each individual's autonomy. With the emergence of central power appeared the first series of texts "for the protection of the public peace." Even after that, the ideology and legislation of representation were only theoretically hostile to noise. Silences reigned in the concerts of the bourgeoisie. Elsewhere, there was no attempt to impose it—for the purpose of making people believe silence is neither possible nor desirable.

The law of December 22, 1789, and article 99 of the law of April 5, 1784, supplemented by the ruling of November 5, 1926 (article 48), provided for only symbolic punishments: article 479 of the French Penal Code that was in force at the beginning of the twentieth century imposed a fine of 11-15 francs for those who disturbed the peace at night while using offensive language. Noise control was the province of the local authorities, who were not very coercive, and was tied to keeping the peace, in other words, monitoring conformity to the norm. Nothing truly repressive was done at that time. In France, the texts were occasionally used to limit the right to assembly in specific cases, always noise related (May 21, 1867, against a dance hall; June 18, 1908, against a concert hall . . .).

The first truly significant campaign against noise in France took place in 1928 on the initiative of dominant social groups. The "Touring Club of France," which organized the campaign, wanted the government to pass comprehensive legislation on industrial and traffic noise. It chose as its motto: "The silence of each assures rest for all." It only succeeded in slightly increasing power's awareness of the benefits of noise control.

Symbolically, noise control was first implemented in relation to an individualized sound object—the automobile. Simultaneously noisemaker, mask, and instrument of death, it is a form of individualized power. The automobile is therefore doubly powerful: the noise it makes is a form of violence, and its camouflage guarantees it impunity. Thus we may risk the hypothesis that the use-level of automobile horns in a city is related to its political and subversive potential, and that the establishment of control over it is indicative of a credible reenforcement of political power at the expense of the subversive elements. The automobile developed with the emplacement of repetition and thus of noise control. Article 25 of the decision of December 31, 1911, which made sounding one's horn a duty, specifies: "However, in densely populated areas, the volume of the sound emitted by the horn should remain low enough that it does not inconvenience the residents and passers-by. The use of multiple-sounding horns, sirens, and whistles is prohibited."

The Highway Code in France was the first in the world. The general police ordinance of February 18, 1948, on traffic control specifies that "all vehicles, with the exception of strollers and carts pulled or pushed by hand, must be equipped with a warning horn that must be used exclusively to warn other vehicles and pedestrians of their approach. This device must have sufficient range and must be capable of being sounded in such a way as to allow the drivers and pedestrians enough time to stand aside or make way.''

A short time later, this control of urban noise had been implemented almost everywhere, or at least in the politically best-controlled cities, where repetition is most advanced.

We see noise reappear, however, in exemplary fashion at certain ritualized moments: in these instances, the horn emerges as a derivative form of violence masked by festival. All we have to do is observe how noise proliferates in echo at such times to get a hint of what the epidemic proliferation of the essential violence can be like. The noise of car horns on New Year's Eve is, to my mind, for the drivers an unconscious substitute for Carnival, itself a substitute for the Dionysian festival preceding the sacrifice. A rare moment, when the hierarchies are masked behind the windshields and a harmless civil war temporarily breaks out throughout the city.

Temporarily. For silence and the centralized monopoly on the emission, audition and surveillance of noise are afterward reimposed. This is an essential control, because if effective it represses the emergence of a new order and a challenge to repetition. It is not essential in representation, which must allow people to speak in order to make them believe. It only becomes essential in repetition.

Do not misunderstand me: controlling noise is not the same as imposing silence in the usual sense. But it is a silence in sound, the innocuous chatter of recuperable cries.

The Theft of Use-Time

Representation stockpiled exchange-time in the form of money. Repetition stockpiles use-time. In repetition, the social demand for a service is expressed more in terms of the possession of an object than its usage: the social demand for music is channeled into a demand for records, the demand for health into medicines. Replicated man finds pleasure in stockpiling the instruments of a deritualized substitute for the sacrifice. There is no longer anything to prompt him to interiorize the act, to experience its fortuitous, vague reality. The absence of noise (of blemish, error) in the stockpiled objects has become a criterion of enjoyment.

Repetition is now under way—the alignment of all production according to a norm, the elimination of any direct relation between the worker and the consumer. This process is at the heart of economic evolution, and it goes a long way back. As we saw above, as early as the time when money replaced barter there has existed a tool in repetition bearing the imprint of power, causing the old modes of exchange to be forgotten, entrapping time; this was the repression of a fundamental human relation, of the enactment of barter. *Just as money con-*

stitutes a stockpile of exchange-time by registering the relative value of things, repetition constitutes a stockpile of use-time by registering their absolute values. At a certain level, accumulation in effect necessitates that one agree to possess, to stockpile usage. It is then necessary to promote the saving of income, abstinence, Lent, as a form of enjoyment. By eliminating use-time after representation had eliminated exchange-time, repetition made possible an explosive growth in production: that growth would have been unthinkable if people had to take the time to negotiate a price when they bought an object that was up for sale, and it would have been quickly impeded if they had to content themselves with producing services at the rate they are consumed. Growth would have been reduced to a combinatorics. Going beyond thus requires stockpiling just as Lent requires penitence. For a society of this kind to survive, it is necessary for people to be able to experience pleasure in conforming to the norm, to repetition, to enslavement, to penitence.

And that is perhaps the key to the process of repetition as it is taking root today. Repetition becomes pleasurable in the same way music becomes repetitive: by hypnotic effect. Today's youth is perhaps in the process of experiencing this fabulous and ultimate channelization of desires: *in a society in which power is so abstract that it can no longer be seized, in which the worst threat people feel is solitude and not alienation, conformity to the norm becomes the pleasure of belonging, and the acceptance of powerlessness takes root in the comfort of repetition.*

The denunciation of "abnormal" people and their usage as innovators[125] is then a necessary phase in the emplacement of repetition. Although training and confinement are the heralds of repetition, confinement is no longer necessary after people have been successfully taught to take pleasure in the norm.

Stockpiling Death

Thus music today is in many respects the monotonous herald of death. Ever since there have musical groups in places where labor consists in dying, death and music have been an indissociable pair. The fact that the musician has always been present at the site of ritual murder, where his role is exceedingly ambiguous, returns us to the essential point about music: sacrifice, music, and the scapegoat form an indestructible whole. The orchestras of Dachau are only the monstrous and modern resurgence of this abominable, everlasting concatenation.

Still today, there is death everywhere in music, and it is no coincidence that many great musicians have chosen physical death (Janis Joplin, Jimi Hendrix, Jim Morrison), or institutional death (the Beatles). Or that theoretical music accepts noise and uncontrolled violence. Or that repetitive music and hyperrealism advertise themselves literally as the murder of creativity, the blasphemous herald of the death of a society in which reality is only a normalized, liquidating artifice. Or that this presence of violence and death can be felt in all of the places of repetition.

For death, more generally, is present in the very structure of the repetitive economy: *the stockpiling of use-time in the commodity object is fundamentally a herald of death.*

In effect, transforming use-time into a stockpileable object makes it possible to sell and stockpile rights to usage without actually using anything, to exchange ad infinitum without extracting pleasure from the object, without experiencing its function.

No human act, no social relation, seems to escape this confinement in the commodity, this passage from usage to stockpiling. Not even the act that is the least separable from use-time: death. Repetition today does indeed seem to be succeeding in trapping death in the object, and accumulating its recording. This is a two-step operation: first, repetition makes death exchangeable, in other words, it represents it, puts it on stage, and sells it as a spectacle. This step is reached not only in films in which the actors are in fact murderers, but also in the American "Suicide Motels," where anyone may choose and purchase an enactment of his own death, and then go through with it. Then in the second step, not yet realized, death will become repeatable, capable of trapping use-time; in other words, purchasing the right to a certain death will become separable from its execution. Thus people will collect means for killing themselves and "death rights," just as they collect records—rights to different deaths (happy, sad, solitary, collective, distant or with the family, painless or after torture, in public or in private). To my mind, this sign will be the ultimate expression of the code of possession. It is inevitable, I think, that this commercialization of death, represented in the commodity and stockpiled in the repetitive economy, will come to pass in the next thirty years. It will signify the definitive emplacement of the society of repetition. This may seem outlandish, unacceptable, and all the more absurd because death is one of the rare operations the use of one form of which excludes using any of the others.

Yet in some sense this stockpiling of death has already been realized; the passage from representation to repetition, from the exchange of death to the entrapment of death-time, has already taken place. But not in the way outlined above. To my mind, this detour explains one of the realities most difficult to explain in the framework of the human sciences: that the nuclear powers have stockpiled the means to destroy the planet several times over. It can only be understood if we interpret it as the stockpiling of death announced above, but collectively rather than individually realized, a route that is both more direct and more tolerable to achieve in repetitive society: it is obviously difficult to lead people individually to take such a leap into the absurd by buying the right to multiple deaths, but it is easy to make them vote for, or to impose upon them, a defense budget to finance those rights, under the pretext of killing others to protect oneself. We can deduce from the preceding discussion that if the "Suicide Motels" were to undergo a rapid development, and if it came to the point that one could buy or sell the rights to use them, the arms race would automatically come to a halt, since individuals would relieve the States of having to make the final extension of the field of repetition, and the collective stockpiling of death would become individualized.

A strange conclusion: in the commodity arena, the only way of halting the arms race today is to promote the sale and private collecting of suicide rights and means.

Everything in our societies today points to the emplacement of the process of repetition. Because death is visible, deafening, because violence is returning, not only in war but in art, in other words, in knowledge, we refuse to take action, to assume it, and to seek a strategy to oppose it. But this is perhaps also because opposing repetition and death requires the courage to speak out on why the classic solutions offered by the economy and politics are ineffective, and in particular the courage to admit that materialism has today become one strategy among others for the funereal emplacement of repetition; the courage to find a way of constructing a political economy entirely different from that of representation, in which death would be accepted for what it is: *an invitation fully to be oneself in life.*

Repetitive Society

Now we are ready to describe the evolution of our society on the basis of its music. We are ready for repetitive society. In the following analysis, there will be neither polemic nor value judgment. We will simply prolong our analysis until it is a caricature of the coming age. An age when death will be everywhere present. When it will be so much of a presence that refusing it will require an urgent effort. That is possible to do. And writing about death is part of that effort—unless this writing is itself a petty element of repetition.

The emplacement of repetitive society can be read today in all of the processes for the production of commodity signs. Generalizing only slightly from our analysis of music and death, we can perhaps bring into view the overall logic of this economic process, so radically different from the preceding one. Analyzing it leads to different models, different sciences, different economic theories, and thus different interpretations of its crisis from those adapted to the crises of the processes of representation, which were crises in the normalization of dissonances.

Crisis is no longer a breakdown, a rupture, as in representation, but a decrease in the efficiency of the production of demand, an excess of repetition. *Metaphorically, it is like cancer, while the crisis of representation is like cardiac arrest.* All of the theoretical perceptions of the contemporary crisis remain incoherent on this point: in the absence of a clear perception of the laws of the political economy of repetition, perceptions of its crisis and inability to legitimate demand are bound to remain mysterious. Echoing the economy of music, I would like to bring out some of the major traits of these *processes of repetition* and of the *crisis of proliferation* that may break out when repetition can no longer function in a stable manner; I would also like to show why the dominant economic theories are unable to account for these.

The Political Economy of Repetition

The repetitive economy is characterized first of all by a mutation in the mode of production of supply, due to the sudden appearance of a new factor in production, *the mold*, which allows the mass reproduction of an original. This fact, so obvious to those who observe our reality, is today still completely neglected by political economy and every kind of social analysis. All of the dominant theories, including Marxism, the critical analysis of representation, continue to reason as though each object was different from the others and was produced by labor that it is possible to isolate in itself. In fact, repetition requires us to reconstruct the essential aspects of the theory, because a given quantity of labor, that of the *molder* who creates the mold, can produce a great number of copies. Therefore, the necessary labor for production is no longer intrinsic in the nature of the object, but a function of the number of objects produced. The information included and transmitted thus plays the role of a stockpile of past labor, of capital.

Molds of this kind are everywhere: computer programs, car designs, medicine formulas, apartment floor plans, etc. The same mutation also transforms the usage of things. The usage to which representative labor was put disappears with mass production. The object replaces it, but loses its personalized, differentiated meaning. A paradox: the object's utility is exchanged for accessibility. Considerable labor must then be expended to give it a meaning, to produce a demand for its repetition.

Repetition is established through the supplanting, by mass production, of every present-day mode of commodity production still inscribed within the network of representation. Mass production, a final form, signifies the repetition of all consumption, individual or collective, the replacement of the restaurant by precooked meals, of custom-made clothes by ready-wear, of the individual house built from personal designs by tract houses based on stereotyped designs, of the politician by the anonymous bureaucrat, of skilled labor by standardized tasks, of the spectacle by recordings of it.

In this network, production is no longer the essential site of creation or competition. Competition takes place earlier, in the creation of the molds, or later, in the production of demand. For the existence of molded objects does not necessarily imply their uniformity or a great number of copies. On the contrary, as in music, repetition requires an attempt to maintain diversity, to produce a meaning for demands.

In the repetitive economy, technological progress is no longer due to individualizable innovations, but to upheavals affecting entire technological systems. For example, in the record industry a major mutation is necessitating an international agreement among the principal producers to further the commercialization of video-disks and players, and to avoid competition between the various possible technologies; another example is the unanimous decision in the auto industry to develop electric cars instead of allowing competition between the various kinds of non-gasoline-powered vehicles.

The essential part of the labor is done outside the production of objects, concerned with producing demand and distributing commodities; the production price of the object becomes a decreasing fraction of the retail sales price, and repetition becomes an essential site for the utilization of unproductive labor. The price system and advertising play only a minor role in producing demand. Since consumer behavior is no longer predictable and no longer takes place within fixed codes, but is on the contrary very volatile and unstable, the importance of marketing can only diminish. The great number of products introduced at low prices and for very narrow markets is what ensures overall commercial success. The consumer dedicates a significant percentage of his time to selecting products introduced almost haphazardly, the usage of which is very difficult to differentiate, except by rankings determined by mysterious processes in which the consumer is led to believe he participates through simulacra of voting.

The economic status of the molders is a variable of cardinal importance in defining the economic organization of the society of repetition. Whether they win the status of authors or authors lose their status, we will see the establishment of a very different social organization. We can foresee, for the highly developed economies, a decentralization of power if their protection is assured, as in music, by a distribution system and molder's associations for the collection of royalties; on the other hand, if they remain salaried employees of private or state-owned concerns, we can foresee a heightening of the centralization of political power.

To my mind, centralization is part of the logic of the growing socialization of royalties in repetition, discussed above. As the mass media share of the royalties rises, the nonsalaried status of the musician will become harder and harder to defend. What will happen is even less predictable in other sectors, where the very notion of the molder is hard to define, since the creative function is spread throughout the production process and use-rights (patents) are held by the organization or the owner of its capital, and not by the inventor.

For the most part, exchange-time and use-time are already incorporated into commodities. The few aspects of life that still remain noncommercialized now (nationality, love, life, death) will in the future become trapped in exchange.

Their spectacle will be put up for sale, but also their accessories, and afterwards their stockpiling. The usage of services (entertainment, health, food) will thus be transformed into a hoard object.

Consumers—a necessary detour in commodity consumption, until it is discovered how to produce them as well—could be replaced by machines to use and destroy production, eliminating human beings once and for all from the repetitive economy they still encumber today. The commodity could also disappear: just as money has become the accountable substitute for dialogue, the commodity could be replaced by the pure sign, a convenient way to stockpile—record jackets; tickets for travel, restaurants, clothes, life, death; passports; love certificates. The political economy of nonsense will have been founded: without man or merchandise.

Finally, repetition must not be confused with stagnation. On the contrary, repetition requires the ongoing destruction of the use-value of earlier repetitions, in other words, the rapid devaluation of past labor and therefore accelerated growth. In *Carnival's Quarrel with Lent*, in which four people pass around pitchers and break them—exchange without usage—Brueghel announces once again that there is crisis after repetition. The process of repetition, in its acceleration, contains the danger of its own downfall.

The Crisis of Proliferation

The crisis of repetition announces a form of crisis different from the one to which we are accustomed in the schemas of representation. Crisis is no longer a breakdown, a rupture. It is no longer dissonance in harmony, but excess in repetition, lowered efficiency in the process of the production of demand, and an explosion of violence in identity. It is far less easy to conceptualize, and much more difficult to circumscribe, than that of representation.

Essentially, proliferation is a manifestation of the difficulty of seeing to it that production is consumed, of giving meaning to commodities, therefore of producing demand apace with the repetitive supply. The process of normalized repetition can in effect only function if commodities are produced at the same time as desires are entrapped and expressed in commodities. If these new needs are slow to appear, if policies—Keynesian or structural—to stimulate consumption fail, production will proliferate without being able to find an outlet; it will repeat itself without being used; it will consequently die from an excess of life, from excessive, uncontrolled *carcinogenic* replication.

In addition, repetition creates identity, therefore rivalry, the first step toward a return of violence. Mimesis eliminates all obstacles to murder, all scapegoats.

The renaissance of violence in our societies, which the pop music of the 1960s so prophetically announced, is the beginning of this crisis of proliferation, by virtue of the silence it implies and the death it announces. Today's violence is not the violence of people separated by a gulf, but rather the final confrontation of copies cut from the same mold who, animated by the same desires, are unable to satisfy them except by mutual extermination.

There are thus two possible strategies: the crisis of proliferation can either be contained, or followed through to the end so a new social order may arise.

In the first strategy, improving the efficiency of the production of demand is seen as the key to the functioning of society. Keynesianism is the babble of this inquiry into the production of demand. It acts on aggregates, but not on the structures of the production of demand; that is, it does not act on the complex and diverse media system, the showcase for products, nor on the overall ideological and concrete process of the production of the consumers themselves. But that is precisely what the focus of future economic policies of repression will have to be. Outside of that, attempting to contain the crisis is to attempt to give a meaning to production, a use-value to the commodity. Today, all progress is thought of as a rehabilitation of use-value, of the durability of products, as a search for new outlets for replicated objects, for greater legitimacy in the definition of consumption or the organization of production, in other words, as an economistic readjustment of the process of repetitive production. The collective appropriation of the means of producing supply and demand thus aids in containing the crisis more than it serves to overturn codes. There is even a high risk that such appropriation will have reactionary results, by making cultural normalization more efficient and broadening the foundation of the repetitive market. Although experience shows that it is actually profit that is sometimes the mainspring of this subversion, that the immediate attraction of profit overpowers the interest in censorship.

This kind of response to the crisis boils down to opening up to all listeners the recorded spectacle of productions which, in the age of representation, were reserved for a minority. This is indeed a way of containing the crisis of proliferation on a long-term basis. But is having Bach or Stockhausen heard by one and all a sufficiently ambitious project to express the consummation of society? Is making the creations of an elite generally available the mark of a blossoming? Must there be an effort to restore the usage of things? Should socialism delay the destruction of commercial codes that capitalism is so good at carrying out? Or would it not be better to allow the general breakup of the old codes to play itself out, so that the conditions for a new language may arise? Even if such a socialism—a reactionary socialism—wanted to, it could not prevent this extermination from continuing, and from eventually reaching the point at which *a break would be made with the pleasure of the simulacrum of usage, in favor of the norm and the stockpiling of signs.*

Already, from within repetition, certain deviations announce a radical challenge to it: the proliferating circulation of pirated recordings, the multiplication of illegal radio stations, the diverted usage of monetary signs as a mode of communicating forbidden political messages—all of these things herald the invention of a radical subversion, a new mode of social structuring, communication that is not restricted to the elite of discourse. When there is an accelerating repetition of the identical, messages become more and more impoverished, and power begins to float in society, just as society floats in music. In representation, power is localized, enacted. Here, it is everywhere, always present, a threatening sound, perpetual listening. In repetitive society, the politician, who with the star is a major incarnation of representative society, loses his role, to the detriment of the institutions of listening and of noise. In the end, the political spectacle itself, now already limited to the highest echelons, may disappear—without, however, a dissolution of power—just as it disappeared in the large corporation, in which legitimacy is now founded upon efficiency and the competence of anonymous, interchangeable cadres, and hardly at all in the personality of the president.

For the second strategy, a new theory of power is necessary. A new politics also: both of these require the elaboration of a politics of noise and, more subtly, a burgeoning of each individual's capacity to create order from noise, outside of the channelization of pleasure into the norm.

Taking a share in power is thus also having one's voice heard. But not necessarily in circumstances of the enactment of power, the function of which is perhaps in the process of disappearing. Literally speaking, "taking power" is no longer possible in a repetitive society, in which the carefully preserved theater of politics is only sustained to mask the dissolution of institutional places of power, to prevent, by perpetuating an illusion, a necessary displacement in the center of gravity of truly subversive and revolutionary acts.

The only possible challenge to repetitive power takes the route of a breach in social repetition and the control of noisemaking. In more day-to-day political terms, it takes the route of the permanent affirmation of the right to be different, an obstinate refusal of the stockpiling of use-time and exchange-time; it is the conquest of the right to make noise, in other words, to create one's own code and work, without advertising its goal in advance; it is the conquest of the right to make the free and revocable choice to interlink with another's code—that is, the right to compose one's life.

Chapter Five
Composing

We see emerging, piecemeal and with the greatest ambiguity, the seeds of a new noise, one exterior to the institutions and customary sites of political conflict. A noise of Festival and Freedom, it may create the conditions for a major discontinuity extending far beyond its field. It may be the essential element in a strategy for the emergence of a truly new society.

In the tumult of time, in the Manichaeism of a political debate stupidly trapped in a facile and sterile economism, opportunities to grasp an aspect of utopia, reality under construction, are too rare not to attempt to use this scanty clue to reconstruct that reality in its totality.

Conceptualizing the coming order on the basis of the designation of the fundamental noise should be the central work of today's researchers. Of the only worthwhile researchers: undisciplined ones. The ones who refuse to answer new questions using only pregiven tools. Music should be a reminder to others that if *Incontri* was not written for a symphony orchestra, or the *Lamentations* for the electric guitar, it is because each instrument, each tool, theoretical or concrete, implies a sound field, a field of knowledge, an imaginable and explorable universe. Today, a new music is on the rise, one that can neither be expressed nor understood using the old tools, a music produced elsewhere and otherwise. It is not that music or the world have become incomprehensible: the concept of comprehension itself has changed; there has been a shift in the locus of the perception of things.

Music was, and still is, a tremendously privileged site for the analysis and revelation of new forms in our society. It announced, before the rest of society, the destruction of sacrifice by exchange and representation, then the stockpiling of the simulacrum of usage in repetition. Thus what once were rites today appear to be wastefulness; what was the foundation of peace appears as antisocial violence; what was an element in the social whole appears as a work of art to be consumed. Our society mimics itself, represents and repeats itself, instead of letting us live.

But the very death of exchange and usage in music, the destruction of all simulacra in accumulation, may be bringing about a renaissance. Complex, vague, recuperated, clumsy attempts to create new status for music—*not a new music, but a new way of making music*—are today radically upsetting everything music has been up to this point. Make no mistake. This is not a return to ritual. Nor to the spectacle. Both are impossible, after the formidable pulverizing effected by the political economy over the past two centuries. No. It is the advent of a radically new form of the insertion of music into communication, one that is overturning all of the concepts of political economy and giving new meaning to the political project. The only radically different course open for knowledge and social reality. The only dimension permitting the escape from ritual dictatorship, the illusion of representation, and the silence of repetition. Music, the ultimate form of production, gives voice to this new emergence, suggesting that we designate it *composition.*

There is no communication possible between men any longer, now that the codes have been destroyed, including even the code of exchange in repetition. We are all condemned to silence—unless we create our own relation with the world and try to tie other people into the meaning we thus create. That is what composing is. Doing solely for the sake of doing, without trying artificially to recreate the old codes in order to reinsert communication into them. Inventing new codes, inventing the message at the same time as the language. Playing for one's own pleasure, which alone can create the conditions for new communication. A concept such as this seems natural in the context of music. But it reaches far beyond that; it relates to the emergence of the free act, self-transcendence, pleasure in being instead of having. I will show that it is at the same time the inevitable result of the pulverization of the networks, without which it cannot come to pass, and a herald of a new form of socialization, for which self-management is only a very partial designation.

Composition is not easy to conceptualize. All political economy up to the present day, even the most radical, has denied its existence and rejected its political organization. Political economy wants to believe, and make others believe, that it is only possible to rearrange the organization of production, that the exteriority of man from his labor is a function of property and is eliminated if one eliminates the master of production. It is necessary to go much further than that. Alienation is not born of production and exchange, nor of property, but of usage: the moment labor has a goal, an aim, a program set out in advance in a code—even if this is by the producer's choice—the producer becomes a stranger to what he produces. He becomes a tool of production, itself an instrument of usage and exchange, until it is pulverized as they are. From the moment there was an operationality to labor, there was exteriority of the laborer. From the moment there was sacrificial ritual coded independently of the musician, the musician lost possession of music. Music then had a goal exterior to the pleasure of its producer, unless he could find pleasure—as is the case in repetition—in his very alienation, in being plugged into codes external to his work, or in his personal recreation of a preestablished score.

Exteriority can only disappear in composition, in which the musician plays primarily for himself, outside any operationality, spectacle, or accumulation of value; when music, extricating itself from the codes of sacrifice, representation, and repetition, emerges as an activity that is an end in itself, that creates its own code at the same time as the work.

Composition thus appears as a negation of the division of roles and labor as constructed by the old codes. Therefore, in the final analysis, to listen to music in the network of composition is to rewrite it: "to put music into operation, to draw it toward an unknown praxis," as Roland Barthes writes in a fine text on Beethoven.[126] The listener is the operator. Composition, then, beyond the realm of music, calls into question the distinction between worker and consumer, between doing and destroying, a fundamental division of roles in all societies in which usage is defined by a code; to compose is to take pleasure in the instruments, the tools of communication, in use-time and exchange-time as lived and no longer as stockpiled.

Is composition future or past? Is there a noise that can organize the transition toward it from the gray world of repetition? Is it possible to read composition in music—if it develops—as an indication of a more general mutation affecting all of the economic and political networks?

Music, in its relation to money, is once again prophetic, announcing the ultimate outcome of the current crisis. Although the excess of repetition heralds a crisis of proliferation, although it renders the production of demand ineffective and the pseudocommunication instituting solitude unacceptable, it also ushers in composition—amid confusion on the part of creators; in and by the death of all the networks; outside codes, exchange, and usage.

It is a foreshadowing of structural mutations, and farther down the road of the emergence of a radically new meaning for labor, as well as new relations among people and between men and commodities. Hear me well: composition is not the same as material abundance, that petit-bourgeois vision of atrophied communism having no other goal than the extension of the bourgeois spectacle to all of the proletariat. It is the individual's conquest of his own body and potentials. It is impossible without material abundance and a certain technological level, but is not reducible to that.

Music is only the first skirmish in a long battle, for which we need a new theory and strategy if we are to analyze its emergence, manifestations, and results. Music is a foretoken of *evolution on the basis of behavior* in the human world, in a crisis announced by artists' refusal to be standardized by money.

The Fracture

Representation made repetition possible by means of the stockpile it constituted. And repetition created the necessary conditions for composition by organizing an amazing increase in the availability of music.

Composition can only emerge from the destruction of the preceding codes. Its beginnings can be seen today, incoherent and fragile, subversive and threatened, in musicians' anxious questioning of repetition, in their works' foreshadowing of the death of the specialist, of the impossibility of the division of labor continuing as a mode of production.

The New Noise

What practice of music should be read as the real harbinger of the future? The pseudonew proliferates today, making it difficult to choose. Musicology always situates this essential fracture back at the entry of noise into music. That was indeed when provocation and blasphemy, the cry and the body, first entered the spectacle. Their entry was imperative in a world in which brutal noise was omnipresent; it did not, however, translate into a real rupture of the existing networks. As early as 1913, Russolo was talking about "the crashing down of metal shop blinds, slamming doors, the hubbub of crowds, the variety of din, from stations, railways, iron foundries, spinning mills, printing works, electric power stations and underground railways and the absolutely new noises of modern war."[127] He invented an orchestra of vibrators, screechers, whistles. Honegger wrote a work, *Pacific 231* (1924), which reproduces the rhythm of the wheels of a train, and Antheil wrote a *Ballet mécanique* (1926), which calls for airplane propellers. In 1929, Prokofiev wrote *Pas d'acier*; Mossolov wrote *The Iron Foundry*; and Carlos Chavez, *HP*.

In Cage, the disruption is more evident; it can be seen in his negation of the channeled nature of music and the very form of the network, in his unconventional use of classical instruments and his contemptuous sneering at the meaning attributed to Art. When Cage opens the door to the concert hall to let the noise of street in, he is regenerating all of music: he is taking it to its culmination. He is blaspheming, criticizing the code and the network. When he sits motionless at the piano for four minutes and thirty-three seconds, letting the audience grow impatient and make noises, he is giving back the right to speak to people who do no want to have it. He is announcing the disappearance of the commercial site of music: music is to be produced not in a temple, not in a hall, not at home, but everywhere; it is to be produced everywhere it is possible to produce it, in whatever way it is wished, by anyone who wants to enjoy it. The composer should

> give up the desire to control sound, clear his mind of music, and set about discovering means to let sounds be themselves rather than vehicles for man-made theories or expressions of human sentiments.[128]

But the musician does not have many ways of practicing this kind of music within the existing networks: the great spectacle of noise is only a spectacle, even if it is blasphemous, or "liquidating," as Roger Caillois said about Picasso. It is not a new code. Both Cage and the Rolling Stones, *Silence* and "Satisfaction," announce a rupture in the process of musical creation, the end of music as an autonomous activity, due to an intensification of lack in the spectacle. They are not the new mode of musical production, but the liquidation of the old.

Announcing the void, voicing insufficience, refusing recuperation—that is blasphemy. But blasphemy is not a plan, any more than noise is a code. Representation and repetition, heralds of lack, are always able to recuperate the energy of the liberatory festival. The Jimi Hendrix Experience inspires dreams, but it does not give one the strength to put its message into practice, to use the musicians' noise to compose one's own order. One participates in a pop music festival only to be totally reduced to the role of an extra in the record or film that finances it. Selling the air in a gallery for a blank check, as Yves Klein did, is not sufficient to produce a new kind of painting, even if in doing so he completed the destruction of exchange.

We must also go beyond the simple reorganization of economic property. For these experiments demonstrate that one cannot get outside the world of repetition simply by attempting to organize the repetitive economy in a new way: the self-management of the repetitive is still repetitive; it is still tied to the same demands for the creation of value and is less efficient, because if the production of supply alone is self-managed, it makes it more difficult to produce demand, which necessitates a linkup with the media. Therefore, attempts to break away from mass music simply by challenging the system of record financing are condemned to failure, unless they are able to transcend themselves, as the example of free jazz will show.

Even if they are a sign of deeper changes and translate basic aspirations, they are not their concretization. Beyond the rupture of the economic conditions of music, composition is revealed as the demand for a truly different system of organization, a network within which a different kind of music and different social relations can arise. A music produced by each individual for himself, for pleasure outside of meaning, usage and exchange.

"Uhuru"[129]*—The Failure of the Economy of Free Music*

Significantly, the economic struggle against repetition began where repetition was born, in the United States, at the center of a struggle against one of the most powerful cultural and economic colonizations. The organized and often consensual theft of black American music provoked the emergence of free jazz, a profound attempt to win creative autonomy, to effect a cultural-economic reappropriation of music by the people for whom it has a meaning.

The lessons of this experiment, of its inability to construct a truly new mode of production, are of capital importance for understanding the problems of composition. They are relevant to the entire political economy of repetition, and are indicative of the difficulties confronting attempts to escape it.

Free jazz was the first attempt to express in economic terms the refusal of the cultural alienation inherent in repetition, to use music to build a new culture. What institutional politics, trapped within representation, could not do, what violence, crushed by counterviolence, could not achieve, free jazz tried to bring about in a gradual way through the production of a new music outside of the industry.

Free jazz, which broke completely with the cautious version of jazz that had gained acceptance, ran into implacable monetary censorship.[130] Certain record companies in the United State went so far as to adopt the policy of no longer recording black musicians, only whites who played like blacks. Thus free jazz very quickly became a reflection of and a forum for the political struggle of blacks in reaction against their insertion into repetition. Attempts were made, rallying all of the colonized forms of music in opposition to the censorship of the official industry, to establish a parallel industry to produce and promote new music.

The musicians of free jazz first organized formally in 1959, when Bill Dixon and Archie Shepp founded the Jazz Composers' Guild, a quasi-union grouping. Then in 1965, the Association for the Advancement of Creative Musicians (AACM) was founded in Chicago; it was a kind of cooperative composed of about thirty black musicians, the purpose of which was to "fight against the dictatorship of the club owners, record companies, and critics" (Archie Shepp). In addition to defending professional interests, its aims were essentially to increase the opportunities for composers, instrumentalists, and groups to meet.

Then, the musicians directed their efforts toward becoming more independent of capital. From this point of view, the most interesting experiment was the Jazz Composers' Orchestra Association, successor of the Jazz Composers' Guild, which had been undermined by internal divisions and racial conflicts.

Founded by a white (Mike Mantler) and several blacks (Carla Bley, etc.), the JCOA encouraged musical research, but tried in particular to develop a distribution and production network for concerts and records parallel to the capitalist circuit. This organization, functioning outside the main structures of the music industry, made it possible for several musicians to work without being obliged to record purely commercial records at too fast a pace.

The association was financed by pooling royalties, and by university and foundation support obtained by music "personalities" and teachers, like Thornton or Shepp. All of the musicians received an equal share of the total earnings. Other original economic organizations were founded in the same spirit, in an attempt to escape the laws of repetition, for example, self-managed orchestras such as Beaver's (360 Degree Experience).

> With the exception of collective action, musicians never do their own promoting. Many blacks who should have been conscious of that a long time ago are realizing it now. We are only just beginning to make our mark and it's going very slowly, because many musicians are penniless. . . . We have to eliminate the studios and all of the middlemen who increase production costs for no reason.[131]

Since repetition today is based essentially on control over distribution, over the production of demand and not the production of the commodity, free jazz ran into difficulty promoting itself from within its own structures, in a world in which repetition monopolizes the major part of the market.

> My recording of *Freedom and Unity* dates from 1967. But I wasn't able to make the record until 1969, and it wasn't until 1971 that the sales paid back the cost of producing it. The aim of a label like Third World is obviously not commercial. It is not so much a matter of selling as launching a collective project giving musicians a chance to progress, with each record having the value of a document of the evolution of our work. . . . To produce *Freedom and Unity*, I worked as a notary public. I was a student at the same time.[132]

A new economic status accompanied the emergence of the new music, a truly spontaneous music of immediate enjoyment that escaped all crystallizations and used instruments in new ways, but was also very carefully crafted and at times very intellectual. The attainment of economic autonomy, the politicization of jazz, and the struggle for integration were contemporaneous. Free jazz, created with the Black Muslims, was experienced as a cultural renaissance, as something new, because of its references to Africa and black pride:

> The white musician can jam if he's got some sheet music in front of him. He can jam on something he's heard jammed before. But that black musician, he picks up his horn and starts blowing some sounds that he never thought of before. He improvises, he creates, it comes from within. It's his soul; it's that soul music. . . . he will improvise; he'll bring it from within himself. And this is what you and I

want. You and I want to create an organization that will give us so much power we can sit and do as we please.[133]

Since that time, the sound of free jazz, like the violent wing of the black movement, has failed in its attempt to break with repetition. It subsided, after being contained, repressed, limited, censored, expelled. After it failed to win real political power, it adopted new forms of creativity and cultural circulation.

The black people of America have adopted a more "reflective" political position, with everything that word implies. . . . It's unwise to take needless risks, to make yourself obvious, to mouth off on television. That makes us vulnerable on every level, easily identified by reactionary forces.[134]

Free jazz, a meeting of black popular music and the more abstract theoretical explorations of European music, eliminated the distinction between popular music and learned music, broke down the repetitive hierarchy.

It also shows how the refusal to go along with the crisis of proliferation created *locally* the conditions for a different model of musical production, a new music. But since this noise was not inscribed on the same level as the messages circulating in the network of repetition, it could not make itself heard. It was the herald of another kind of music, a mode of production outside repetition— after having failed as a *takeover of power in repetitive society.*

Representation and Composition: The Return of the Jongleurs

Along with this search for another power in repetition, continuing this mutation in the economy of music, there is a reappearance of very ancient forms of production.

First, there is a resurgence in the production of popular music using traditional instruments, which often are handmade by the musicians themselves—a resurgence of music for immediate enjoyment, for daily communication, rather than for a confined spectacle. No study is required to play this kind of music, which is orally transmitted and largely improvisational. It is thus accessible to everyone, breaking the barrier raised by an apprenticeship in the code and the instrument. It has developed among all social classes, but in particular among those most oppressed (the workers of the big industrial cities, Black American ghettos, Jamaican shantytowns, Greek neighborhoods, etc.). "True creativity lies with the foreigners, and culture is on the side of those who live on the margins of culture without living with it . . . the metics" (O. Revault d'Allones). Their music is generally without cultural references.

The number of small orchestras of amateurs who play for free has mushroomed. Music is thus becoming a daily adventure and an element of the subversive festival again. A very significant fact: the production and invention of instruments, nearly interrupted for three centuries, is noticeably increasing. The creative labor is collective: what is played is not the work of a single creator; even if an individual's composition is taken as the point of departure, each musician develops his own instrumental part. Production takes the form of one of collective composition, without a predetermined program imposed upon the players, and without commercialization. The groups stay together only a short time, dissolving when their members rejoin repetitive life. This music is creating a new practice of musical production, a day-to-day and subversive practice. It is incontestably aided by the existence of the records and representations it subverts, from which it draws inspiration and innovations. A new usage of records is also developing: records are being made with only instrumentation, meant to be sung to, in other words, allowing one to insinuate oneself into production (*Minus One*).

This does not constitute, therefore, a new form of popular music, but rather a new practice of music among the people. Music becomes the superfluous, the unfinished, the relational. It even ceases to be a product separable from its author. It is inscribed within a new practice of value. The labor of music is then essentially an "idleness" (D. Charles) irreducible to representation (to exchange) or to repetition (to stockpiling). It heralds the negation of the tool-oriented usage of things. By subverting objects, it heralds a new form of the collective imaginary, a reconciliation between work and play.

Although these new practices may faintly resemble those of the medieval jongleurs, they in fact constitute a break with sacrificial, representative, and repetitive music: before the advent of recording and modern sound tools, the jongleurs were the collective memory, the essential site of cultural creation, and the circulation of information from the courts to the people. Recording stabilized the musical work and organized its commercial stockpiling. But now the field of the commodity has been shattered and a direct relation between man and his milieu is being reestablished.

Music is no longer made to be represented or stockpiled, but for participation in collective play, in an ongoing quest for new, immediate communication, without ritual and always unstable. It becomes nonreproducible, irreversible. "If we compose music, we are also composed by history, by situations that constantly challenge us" (L. Berio). Music is ushering in a new age. Should we read this emergence as the herald of a liberation from exchange-value, or only of the emplacement of a new trap for music and its consumers, that of automanipulation? The answer to these questions, I think, depends on the radicality of the experiment. Inducing people to compose using predefined instruments cannot lead to a mode of production different from that authorized by those instruments.

That is the trap. The trap of false liberation through the distribution to each individual of the instruments of his own alienation, tools for self-sacrifice, both monitoring and monitored. A trap upon which singing to recorded music and

Minus One shed direct light, since their future development will permit their inscription within a code, their becoming a form of semiautonomous participation, a tool for liberation and a pedagogy of channelization.

The Relational Value of Composition

The New Unity of the Body

Thus in a reversal of the current process, which starts with the conception and ends with the object, the outcome of labor no longer "pre-exists ideally in the imagination of the worker."[135] To modify the meaning of form in the course of its production, to empty exchange/use-value of its alienating content, is to attempt to designate the unsayable and the unpredictable. A structuring of desire and production can then be outlined on the basis of what music makes audible of them.

The essential mutation, of course, is in the relation to oneself that music makes possible. The disappearance of codes, and the destruction of the communication that took place in the sacrifice or the commercial simulacrum, at first open the way for the worker's reappropriation of his work. Not the recuperation of the product of his labor, but of his labor itself—labor to be enjoyed in its own right, its time experienced, rather than labor performed for the sake of using or exchanging its outcome. The goal of labor is no longer necessarily communication with an audience, usage by a consumer, even if they remain a possibility in the musical act of composition. The nature of production changes; the music a person likes to hear is not necessarily the same music he likes to play, much less improvise. In composition—the absence of exchange, self-communication, self-knowledge, nonexchange, self-valorization—labor is not confined within a preset program. There is a collective questioning of the goal of labor. To my knowledge, the economic organization of this form of production lacking defined goals, and the nature of the new relation it creates between man and matter, consumption-production and pleasure, have never been expressed in theory before.

In composition, to produce is first of all to take pleasure in the production of differences. Musicians foresaw this new concept. For example, in the language of jazz, to improvise is "to freak freely." A freak is also a monster, a marginal. To improvise, to compose, is thus related to the idea of the assumption of differences, of the rediscovery and blossoming of the body. "Something that lets me find my own rhythm between the measures" (Stockhausen). Composition ties music to gesture, whose natural support it is; it plugs music into the noises of life and the body, whose movement it fuels. It is thus laden with risk, disquieting, an unstable challenging, an anarchic and ominous festival, like a Carnival with an unpredictable outcome. This new mode of production thus entertains a very different relation with violence: in composition, noise is still a metaphor for murder. To compose is simultaneously to commit a murder and to perform a sacrifice. It is to become both the sacrificer and the victim, to make an ever-possible suicide the only possible form of death and the production of life. To compose is to stay repetition and the death inherent in it, in other words, to locate liberation not in a faraway future, either sacred or material, but in the present, in production and in one's own enjoyment.

Relationship

Composition does not prohibit communication. It changes the rules. It makes it a collective creation, rather than an exchange of coded messages. To express oneself is to create a code, or to plug into a code in the process of being elaborated by the other.

Composition—a labor on sounds, without a grammar, without a directing thought, a pretext for festival, in search of thoughts—is no longer a central network, an unavoidable monologue, becoming instead a real potential for relationship. It gives voice to the fact that rhythms and sounds are the supreme mode of relation between bodies once the screens of the symbolic, usage and exchange are shattered. In composition, therefore, music emerges as a relation to the body and as transcendence.

Of course, in sacrifice and representation, the body is already contained in music: Ulysses' companions risked dying of pleasure by listening to the song of sirens, and the duos in *Cosi Fan Tutte* and *Tristan und Isolde* express a real erotic drive. Music, directly transected by desires and drives, has always had but one subject—the body, which it offers a complete journey through pleasure, with a beginning and an end. A great musical work is always a model of amorous relations, a model of relations with the other, of eternally recommenceable exaltation and appeasement, an exceptional figure of represented or repeated sexual relations. "With music, I am almost incapable of obtaining any pleasure," wrote Freud;[136] and yet psychoanalysis has had next to nothing to say about noise and music.

But in composition, it is no longer, as in representation, a question of marking the body; nor is it a question of producing it, as in repetition. It is a question of taking pleasure in it. That is what relationship tends toward. An exchange between bodies—through work, not through objects. This constitutes the most fundamental subversion we have outlined: to stockpile wealth no longer, to transcend it, to play for the other and by the other, to exchange the noises of bodies, to hear the noises of others in exchange for one's own, to create, in common, the code within which communication will take place. The aleatory then rejoins order. Any noise, when two people decide to invest their imaginary and their desire in it, becomes a potential relationship, future order.

From Noise to Image: The Technology of Composition

Like representation and repetition, composition needs its own technology as a basis of support for the new form of value. While recording was intended as a reinforcement for representation, it created an economy of repetition. As with the preceding codes, the technology upon which composition is based was not conceived for that purpose. If representation is tied to printing (by which the score is produced), and repetition to recording (by which the record is produced), composition is tied to the instrument (by which music is produced). We may take this as a herald of considerable future progress, in the production and in the invention of new instruments.

Once again, music appears to me to be premonitory. The current burgeoning of instruments in the expansive field of sound—as great as that of the sixteenth and seventeenth centuries, which announced the industrial revolution—foreshadows a new mutation in technology.

In this context, there is an innovation that is only now beginning to play out its role, a herald of this mutation: the recording of images. Today, the recording of images is intended to be an instrument for the visual stockpiling of concerts and films and as a means of pedagogy, in other words, as a tool of repetition. Soon, however, it may become one of the essential technologies of composition. Television, the prehistory of image recording, did not succeed in giving visual status to music; the body disappeared, and individual image recording emerged as an innovation devoid of music. In a preliminary period, it may become a means of stockpiling access to films, on an individual basis or in the form of a central memory bank. But the essential usage of the image recorder seems to me to be elsewhere, in its private use for the manufacture of one's own gaze upon the world, and first and foremost upon oneself. Pleasure tied to the self-directed gaze: Narcissus after Echo. Eroticism as an appropriation of the body.

The new instrument thus emerging will find its real usage only in the production, by the consumer himself, of the final object, the movie made from virgin film. The consumer, completing the mutation that began with the tape recorder and photography, will thus become a producer and will derive at least as much of his satisfaction from the manufacturing process itself as from the object he produces. He will institute the spectacle of himself as the supreme usage.

The Political Economy of Composition

Composition belongs to a political economy that it is difficult to conceptualize: production melds with consumption, and violence is not channeled into an object, but invested in the act of doing, a substitute for the stockpiling of labor that simulates sacrifice. Each production-consumption (composition) entity can call its program into question at any moment; production is not foreseeable before its conclusion. It becomes a starting point, rather than being an end product; and time is lived time, not only in exchange and usage, but also in production itself. The bulk of commodity production then shifts to the production of tools allowing people to create the conditions for taking pleasure in the act of composing. We can see—in the makeup of musical groups, in the creation of new instruments, in the development of the imaginary through the planning of personal gardens,[137] in production using rudimentary tools—an outline of what composition can mean: each person dreaming up his own criteria, and at the same time his way of conforming to them.

Just as the enjoyment of music no longer passes through exchange or stockpiling, the enjoyment of production is exterior to its insertion in a market or system of allocation. It is thus necessary to conceive of other systems of economic organization, and especially other political institutions. For violence is no longer channeled into sacrifice; it no longer mimics itself in representation; it is no longer threatening, as it was in repetition. The wager of the economy of composition, then, is that social coherence is possible when each person assumes violence and the imaginary individually, through the pleasure of doing.

Composition liberates time so that it can be lived, not stockpiled. *It is thus measured by the magnitude of the time lived by men, which takes the place time stockpiled in commodities.*

One may wonder whether a model such as this, composed of liberated time and egoistic enjoyment, is possible. And in fact, on closer inspection, seemingly insoluble problems of coherence arise: first, others' noise can create a sound of cacophony, and each difference thus created, between units of composition, may be felt as a nuisance. Second, the complementarity of productions is no longer guaranteed, because compositional choices are not confronted by a pricing system (the market in representation) or ranking (planning in repetition).

Thus this social form for the recreation of difference—assuming it does not fall back into the commodity and its rules, in other words, into representation and repetition—presupposes the coexistence of two conditions: *tolerance and autonomy.* The acceptance of other people, and the ability to do without them. That being the case, composition obviously appears as an abstract utopia, a polar mode of organization that takes on meaning at an extraordinary moment of cultural climax.

Even then, composition may be considered an impossibility. There are several reasons for this. First, as Pierre Boulez writes:

It is necessary to deny all invention that takes place in the framework of writing. . . . Finally, improvisation is not possible. Even in a baroque ensemble, where the laws were more or less codified, where you had figures instead of chords, in other words, where you could place them in a certain position but not in just any way—even in this period improvisation did not produce exclusively masterpieces. People speak of Bach's improvisations, for example. I believe that Bach wrote on the basis of what he had improvised, and that what he wrote was the more interesting of the two. Often, these improvisations are nothing more than pure, sometimes bizarre, samplings of sound that are not at all integrated into the directives of a composition. This results in constant arousal and appeasement, something I find intolerable. . . . The dialectic of form takes precedence over the possible; everybody arouses everybody else; it becomes a kind of public onanism.

The impossibility of improvisation thus forbids composition. The second reason is given by Claude Lévi-Strauss, who writes that it is difficult to concede that there exists in each person the potential for musical creation:

Theoretically, if not in fact, any adequately educated man could write poems, good or bad; whereas musical invention depends on special gifts, which can be developed only where they are innate.[138]

If both Boulez and Lévi-Strauss, or either one, are right there will be no composition. But nothing in modern biology confirms the validity of these value statements, which confuse creativity with the present code of creativity. Neither will there be composition if it is not clearly willed as a project to transcend repetition, in other words, if the State does not stop confusing well-being with the production of demand. Any policy that valorizes the usage of objects instead of the means of producing them retards composition. On the other hand, a massive decentralization of power would accelerate it. In this case, then, the transition from one network to another is very different from the two preceding transitions. For the first time, it is not in the interest of the economic apparatus. For the first time, the requirements of the accumulation of commodity-value are reactionary and demand policies that are objectively conservative, even if they are camouflaged as an equalization of the conditions of access to commodities. In this sense, the creators themselves are in a precarious position, because composition contains the germ of their disappearance as specialists. So what noise will arrive to create the new order? We have seen that musicians' regaining their music is not enough. There is only one way: recovering—in the units of production and of life, in undertakings and collectivities—some meaning for things. The State can play a positive role only by encouraging the extensive production of means of doing rather than objects, the production of instruments rather than music. A profound mutation, delocalized and diffuse, that fundamentally changes the code of social reproduction, thus leading to a radical challenge to the somber power of the managers of repetition.

But the dangers are immense, for once the repetitive world is left behind, we enter a realm of fantastic insecurity.

Music no longer recounts a mastered, reasoned history. It is inscribed in a labyrinth, a time graph. After the third stage of the attainment of power described by Castaneda has been passed, the stage in which man has conquered power, the relation to technology and knowledge changes, because the relation to the essential has changed. Three moments interpenetrate and stand in opposition to one another.

Perhaps the reader will have remarked what mysterious and powerful links exist between technology and knowledge on the one hand, and music on the other. Everywhere present, lurking behind a form, knowledge molds itself to the network within which it is inscribed: in representation, it is a model, a schema, the value of which depends on its empirical suitability to the measurement of facts; it is the study of partitions (*partitions*, also "scores"). In repetition, it is genealogy, the study of replication. In composition, it is cartography, local knowledge, the insertion of culture into production and a general availability of new tools and instruments.

Composition thus leads to a staggering conception of history, a history that is open, unstable, in which labor no longer advances accumulation, in which the object is no longer a stockpiling of lack, in which music effects a reappropriation of time and space. Time no longer flows in a linear fashion; sometimes it crystallizes in stable codes in which everyone's composition is compatible, sometimes in a multifaceted time in which rhythms, styles, and codes diverge, interdependencies become more burdensome, and rules dissolve.

In composition, stability, in other words, differences, are perpetually called into question. Composition is inscribed not in a repetitive world, but in the permanent fragility of meaning after the disappearance of usage and exchange. It is neither a wish nor an anxiety, but the future contained in the history of the economy and in the predictive reality of music. It is already present—in its fragility and instability, in its transcendence and fortuitousness, in its requirement of tolerance and autonomy, in its estrangement from the commodity and materiability—implicit in our everyday relation to music. It is also the only utopia that is not a mask for pessimism, the only Carnival that is not a Lenten ruse.

It announces something that is perhaps the most difficult thing to accept: henceforth *there will be no more society without lack*, for the commodity is absolutely incapable of filling the void it created by suppressing ritual sacrifice, by deritualizing usage, by pulverizing all meaning, by obliging man to communicate first to himself.

Living in the void means admitting the constant presence of the potential for revolution, music and death: "What can a poor boy do, except play for a rock 'n' roll band?" ("Street Fighting Man," Rolling Stones). Truly revolutionary music is not music which expresses the revolution in words, but which speaks of it as a lack.

Bringing an end to repetition, transforming the world into an art form and life into a shifting pleasure. Will a sacrifice be necessary? Hurry up with it, because—if we are still within earshot—the World, by repeating itself, is dissolving into Noise and Violence.

Five people in a circle. Are they singing? Is there an instrument accompanying them? Is Brueghel announcing this autonomous and tolerant world, at once turned in on itself and in unity?

For my own part, I would like to hear the Round Dance in the background of *Carnival's Quarrel with Lent* as the culmination, not the inauguration, of a struggle begun twenty-five centuries ago. I would like to hear it as the forerunner of postpenitence, postsilence, at the back exit of the church, not the rearguard of the pagan Carnival, supplanted by capitalist Lent in the foreground.

Unless Brueghel, by making the field interpenetrate, rooting each within the other, wishes to signify that everything remains possible and to make audible, as though by a message coded in irony, the inevitable victory of the aleatory and the unfinished.

Afterword
The Politics of Silence and Sound
Susan McClary

The subject of Attali's book is noise, and his method is likewise noise. His unconcealed ideological premises, his penchant for sullying the purity of pitch structures with references to violence, death, and (worst of all) money, and his radically different account of the history of Western music all jar cacophonously against the neat ordering of institutionalized music scholarship, especially as it is practiced in the United States. It is, therefore, quite conceivable that those trained in music will perceive the book's content also as noise—that is, as nonsense—and dismiss it out of hand.

Such dismissal would not surprise Attali, for among his observations he includes remarks on the rise of positivistic musicology and pseudoscientific music theory, both of which depend upon and reenforce the concept that music is autonomous, unrelated to the turbulence of the outside, social world. But it would be most unfortunate if the mechanisms that have already done so much to silence the human and social dimensions of our music (past and present, classical and popular) succeeded also in silencing the noise of this book. For if Attali can serve to jolt a few musicians awake or to encourage those attempting to forge new compositional or interpretive directions, then the hope he expresses for a new music—controlled neither by academic institutions nor by the entertainment/recording industry—may be at least partially realized.

Noise poses so many provocative questions that to try to respond adequately to it would require another book—or, indeed, new fields of study, new modes of creating, distributing, and listening to music. In this essay I shall address and amplify three issues raised by Attali: first, the means by which silence has been imposed and is maintained by our theories and histories of music, by our performance practices and educational institutions; second, the concept that music articulates the ways in which societies channel violence and some ways in which this concept might be used in constructing a revised history of music; and third, the most recent of Attali's four stages of music (Composition) and signs of its emergence in the seven years that have passed since the original publication of *Noise*.

The idea that music can be regarded as silenced, even as its din surrounds us deafeningly at all times, seems a paradox, but it is central to Attali's argument. Unless one can accept this idea and its far-reaching implications, one cannot respond sympathetically to his narrative or prognosis. But the theories of music that have shaped our perceptions and consumption of music have been instrumental in conditioning us not to recognize silencing—not to realize that something vital may be missing from our experience.

From the time of the ancient Greeks, music theory has hovered indecisively between defining music as belonging with the sciences and mathematics or with the arts. Its use in communal rituals and its affective qualities would seem to place it among the products of human culture, yet the ability of mathematics to account for at least some of its raw materials (tones, intervals, etc.) has encouraged theorists repeatedly to ignore or even deny the social foundations of music. The tendency to deal with music by means of acoustics, mathematics, or mechanistic models preserves its mystery (accessible only to a trained priesthood), lends it higher prestige in a culture that values quantifiable knowledge over mere expression, and conceals the ideological basis of its conventions and repertories. This tendency permits music to claim to be the result not of human endeavor but of rules existing independent of humankind. Depending on the conditions surrounding the production of such a theory, these rules may be ascribed to the physical-acoustical universe or may be cited as evidence for a metaphysical realm more real than the imperfect material, social world we inhabit.

Now it is quite clear to most listeners that music moves them, that they respond deeply to music in a variety of ways, even though in our society they are told that they cannot know anything about music without having absorbed the whole theoretical apparatus necessary for music specialization. But to learn this apparatus is to learn to renounce one's responses, to discover that the musical phenomenon is to be understood mechanistically, mathematically. Thus nontrained listeners are prevented from talking about social and expressive dimensions of music (for they lack the vocabulary to refer to its parts) and so are trained musicians (for they have been taught, in learning the proper vocabulary, that music is strictly self-contained structure). Silence in the midst of sound.

A few examples. Jean Philippe Rameau is recognized as the founder of tonal harmonic theory—the theory developed first to account for music of the eighteenth century, later extended to nineteenth-century repertories. Musicians have been trained for the last two hundred years to perceive music in Rameau's terms —as sequences of chords—and thus his formulations seem to us self-evident. Before Rameau's *Traité de l'harmonie* [*Treatise on Harmony*] (1722), theories and pedagogical methods dealt principally with two aspects of music: coherence over time (mode) and the channeling of noise in the coordination of polyphonic voices (counterpoint). In this tradition, the integrity of a composition's sense of motion and formal unfolding was preserved, and simultaneities were treated contextually—as formations that emerged from communal activity and that continued on in accordance with rules for dissonance control, with the verbal text, and with the modal structure. Rameau, in a striking reworking of Descartes' *Cogito* manifesto, declared this earlier tradition moribund and, in seeking to build a musical system from reason and science, hailed the triad as the basis of music.

Now to be sure, the major triad can be generated from very simple mathematical principles, and its pitches occur in the overtone series. It appears thus to be inscribed in nature (not invented arbitrarily by culture), and its music seems to be therefore the music dictated by the very laws of physics. Yet the triad is inert. Breaking a piece of music down into a series of its smallest atomic units destroys whatever illusion of motion it might have had. It yields a chain of freeze-frame stills, all of which turn out to be instances of triads. Mathematical certainty and the acoustical seal of approval are bought at the price of silence and death, for text, continuity, color, inflection, expression, and social function are no longer relevant issues. The piece is paralyzed, laid out like a cadaver, dismembered, and cast aside.

Heinrich Schenker's neo-Hegelian theoretical program early in this century attempted to restore to music theory the accounting for motion, the illusion of organically unfolding life he detected in German music from Bach to Brahms. His principal treatise, *Der freie Satz* [*Free Composition*] (1935), is expressly metaphysical in intent—the work of an Austrian Jew between the world wars who sought evidence of transcendental certainty and meaning in this music. The book is intensely, almost desperately, rigorous as he demonstrates the underlying process that characterizes all "great" (that is, eighteenth and nineteenth-century German) music. Details of expression, rhetoric—even vocal texts—are dismissed as surface irrelevances in his search for higher truth. Ironically, while his treatise provides the key to much of the implicit ideology of the standard German repertory, Schenker conceals his observations in formalisms. As a further irony, Schenker's work has been accepted as one of the principal modes of academic analysis in the United States, but only after it was stripped of its ideological trappings: in the recent translation (trans. Ernst Oster [New York:Longman, 1979]) the sections involved with mysticism and German supremacy have been moved to an appendix. The book now reads like a cut-and-dried method and is meant to be used as one. If Schenker silenced the cries of uncertainty and anguish apparent in the discontinuities of so much nineteenth-century music by showing that it all is—in the final analysis—normative and consistent with the laws of God, American Schenkerians have in turn silenced his metaphysical quest.

What does it matter in the real world of production and listening what music theorists say to one another? Inasmuch as musicians who are trained in conservatories or universities are required to have had at least two (often three or four) years of such theoretical study, it can matter quite a bit: our performers, historians, and composers by and large are taught that music has no meaning other than its harmonic and formal structure.

The performers on whom we rely to flesh out notated scores into sound are trained *not* to interpret (understood as the imposition of the unwanted self on what is fantasized to be a direct transmission of the composer's subjective intentions to the listener), but rather to strive for a perfect, standard sound, for an unbroken, polished surface. Such performers became ideal in the nineteenth century as grist for the symphony orchestra in which the conductor usurped complete control over interpretation and needed only the assurance of dependable sound production from the laboring musician. In our century of Repetition, they have remained ideal for purposes of the recording industry, which demands perfection and the kind of consistency that facilitates splicing. And our mode of consuming music as background decor (Beethoven's C# Minor Quartet played as Muzak at academic cocktail parties) favors performances that call no attention to themselves.

Because Attali's book locates musical social significance in its channeling of noise and violence—qualities almost entirely lacking in our musical experiences—his point is likely to be met with incomprehension. But he is absolutely right. If the noise of classical music (portrayals of the irrational in Bach, the Promethean struggles of Beethoven, the bitter irony and agonizing doubt of Mahler) is no longer audible, it is because it has been contained by a higher act of violence. To refuse to enact the ruptures of a discontinuous musical surface is to silence forcibly, to stifle the human voice, to render docile by means of lobotomy. It is this mode of performance that characterizes our concert halls and recordings today. It leads us to believe that there never was meaning, that music always has been nothing but pretty, orderly sound.

Likewise historians of music, given their commitment to positivistic research and formal descriptions of music, limit their programs to questions that can be answered factually. Problems of the sort Attali raises are not simply solved differently in musicology—they are not even posed, for to attempt solving them would lead necessarily into forbidden speculation. If the piece of music is but a series of chords on a notated grid, then there exists no way of linking it to the outside world. Research involves the conditions surrounding the material production of the work and the preparation of increasingly rigorous scholarly editions. Musicology remains innocent of its own ideology, of the tenets with which it marks the boundaries between its value-free laboratory and the chaotic social world. Reduced to an artifact to be dated and normatively described, the piece of music is sealed and stockpiled, prevented from speaking its narrative of violence and order.

Composers raised within the academic context have been silenced in a way perhaps more detrimental than other members of the musical caste. For the music of the concert repertory (the mainstay of performers, musicologists, and theorists) did at least get to present some semblance of live drama at some time in history. But the university that has provided a shelter for alienated artists for the last forty years has also encouraged them to pursue increasingly abstract, mathematically based, deliberately inaccessible modes of composition. A curious reversal has occurred: the relentless serial noise of Schoenberg's protest against the complacent bourgeoisie has become the seat of institutionalized order, while attempts by younger composers to communicate, to become expressive, are dismissed as noise—the noise of human emotion and social response. The battle between the New York Uptown and Downtown schools of composition (which will be dealt with again later in this essay) is being waged over what counts as noise, what counts as order, and who gets to marginalize whom. Attali's *Noise*, as it traces the contours of the invisible, inaudible network controlling our musical world, helps immeasurably in clarifying the issues underlying today's upheavals.

Attali's model for the ideological criticism of music (based on the idea that the relationship between noise and order in a piece or repertory indicates much about how the society that produced this music channels violence) owes a great deal to Theodor Adorno. Attali's model differs, however, in that Attali is not bound up with Adorno's love/hate relationship with German culture, which caused him on the one hand to despise all else as trivial or primitive, but on the other to call attention to signs of totalitarianism, self-willed silence, and finally death in the German musical tradition. Adorno's program is first that of a Cassandra and then that of a coroner performing an autopsy. Attali may likewise resemble Cassandra (and the future may prove him a coroner as well), but his model permits him to consider a much wider spectrum of music, to recognize the German tradition as an extremely important moment in the continuum of Western music, but to be able in addition to recognize popular genres and ethnic, early, and new musics. The insights of both Adorno and Attali, however, are results of a refusal to read the history of music as a flat, autonomous chronological record, an insistence on understanding musical culture of the past as a way of grasping social practices of the present and future. Both take the music we retreat to for escapist fantasies or entertainment and convert it into discomforting reminders of that from which we sought to hide—political control and money.

If American musicology is concerned with polishing the surfaces of compositions for affirmative appreciation—indeed with polishing the entire history of style into a chain of bright, attractively packaged commodities—what sort of historical narrative would Attali's model produce? He has provided an outline, filled in occasionally with evocative examples that whet (but do not fully satisfy) the appetite. Such spottiness is characteristic of the early stages of most paradigm shifts. But his model does offer the key to a revitalized version of the history of Western (and even, by extension, non-Western) music, and it is possible to apply it productively to repertories he does not discuss at length.

For instance, several elements of seventeenth-century music can be richly illuminated both by Attali's succession of stages (Sacrifice, Representation, etc.) and by his concept of examining the opposition of socially legitimated order and noise in explaining style change. Polemic discussions concerning style—rival taxonomies, competitive claims to authorized lines of descent, and ideologically polarized sets of tastes—were rampant in the seventeenth century, indicating that it might be a period of particular interest to an enterprise connecting music and social/economic factors. But the seventeenth century is not usually treated very seriously in musicology, for its music is (in terms of our standard tonal expectations) noise. If we take Attali at his most daring and permit ourselves to assume that music truly heralds changes that are only later apparent in other aspects of culture, we may find explanations for several problems in seventeenth-century music scholarship: for the upheaval in style around 1600, for the peculiar contradictory story concerning the invention of opera advanced by its first practitioners, for the staunch resistance in France to Italian style, and for modern musicology's tendency to write the century off as primitive.

Attali locates the stage of Representation (music for the bourgeois audience) in the nineteenth century. I wish to propose that it appeared much earlier, that it was ushered in with great fanfare with the invention of opera, monody, and sonata in the first decade of the seventeenth century.

Indeed, opera was first called *stile rappresentativo*, and its express purpose was to make spectators believe in—to experience directly—the dramatic struggles enacted in its performances. In place of the equal-voiced polyphony of the previous style (now dubbed by the rebels as the *prima prattica* ["first practice," as opposed to the new, modern "second practice"], it made use of flamboyant, virtuosic individuals. Its technical means involved a particular transformation of earlier syntactical procedures that resulted in constant surface control and long-term goal orientation (the essential ingredients of tonality and, not coincidentally, of capitalism).

It is significant that opera (and parallel solo genres) developed not in the context of the hereditary feudal aristocracy (which is often assumed by historians and social critics of opera), but in the courts of northern Italy that were sustained by commerce and later, after 1637, in public opera houses. Despite the humanistic red herrings proffered by Peri, Caccini, and others to the effect that they were reviving Greek performance practices, these gentlemen knew very well that they were basing their new reciting style on the improvisatory practices of contemporary popular music. Thus the eagerness with which the humanist myth was constructed and elaborated sought both to conceal the vulgar origins of its techniques and to flatter the erudition of its cultivated patrons.

Moreover, the plots themselves repeatedly involve the subversion of the inherited social hierarchy. Orfeo as a demigod (between the gods and the plebian shepherds) willfully breaks through traditional barriers, first in his seduction of the deified nobility through his great individual virtuosity that wins him admission into the forbidden Inferno, and second in his apotheosis. Monteverdi's Poppea, Alidoro in Cesti's *Orontea*, and Scarlatti's Griselda succeed in penetrating the aristocracy by force of their erotic charms, talent, or virtue (all of which qualify as noise in a static, ordered social structure). What is represented, what one is made to believe in this music is the rightful emergence of the vital, superior middle-class individual in defiance of the established, hereditary class system.

That there should have been attempts at dismissing the new style as noise is to be expected, and the spokesmen for traditional authority rushed in with lists of errors committed by the new composers in voice-leading and dissonance control (quite literally complaints concerning the mischanneling of violence). The almost raw erotic energy of the new style swept over Europe, nonetheless, meeting real opposition in only one place: the France of Louis XIV. This too is to be expected, for the individual-centered explosivity of the Italian compositional procedures (with their compelling momentum, enjambments, and climaxes), performance practices (with improvised effusions added on the spot by the individual singer), and subversive plots could only have revealed the oppressiveness of Louis' absolutist regime of enforced Platonic harmony. Italian music was, in fact, banned in France, clearly for ideological reasons; but the documents comparing Italian and French styles refer not to politics directly, but to matters of orderliness, harmoniousness, and tastefulness (French *bon goût* or good taste versus Italian noise). If the violence of Italian music is right on the surface, luring us along and detonating periodically to release its pent-up tension, violence is equally present in French music—but it is inaudible. It is that which has silenced the noise, systematically siphoned off the tension, leaving only pretty blandness. The most worrisome aspects of music to a regimented society—the areas in which noise is most likely to creep in, such as physical motion and ornamentation—are the most carefully policed in French performance. Exact formulas for the bowing of stringed instruments and for the precise execution of ornaments were codified and enforced: the performer was most regulated exactly where he would ordinarily be permitted to exercise greatest individual spontaneity.

Why does musicology avoid taking the seventeenth century seriously? Precisely because the ideological struggles that put tonality, opera, and solo instrumental music (and their economic, philosophical, and political counterparts) in place by the eighteenth century are distressing to witness—especially if one wants to hang onto the belief that tonality (and capitalism, parliamentary democracy, Enlightenment rationalism) are inevitable and universal. Only when the dust of the seventeenth century settles and the new ideological structures are sufficiently stabilized to seem eternal can we begin to perform acts of canonization and the kind of analysis that seeks to confirm that ours is truly the only world that works. The seventeenth century reveals the social nature and thus the relative status of tonal music's "value-free" foundation.

This interpretation of the seventeenth century goes counter to Attali's only in that he places the transition to Representation considerably later. It validates, however, the concepts central to his position: that music announces changes that only later are manifested in the rest of culture and that it is in terms of the noise/order polarity that styles define themselves ideologically against predecessors or contemporaneous rival practices. A history of Western music rewritten on the basis of these principles would be extraordinarily valuable, for musicians still stuck with sterile chronologies, but especially for nonmusicians who (as Attali demonstrates so well) must have access to the kinds of insights music offers.

Attali's term for the hope of the future, *Composition*, seems strange at first glance, for this is the word used in Western culture for centuries to designate the creation of music in general. But the word has been mystified since the nineteenth century, such that it summons up the figure of a semidivine being, struck by holy inspiration, and delivering forth ineffable delphic utterances. Attali's usage returns us to the literal components of the word, which quite simply means "to put together." It is this demystified yet humanly dignified activity that Attali wishes to remove from the rigid institutions of specialized musical training in order to return it to all members of society. For in Attali's eyes, it is only if the individuals in society choose to reappropriate the means of producing art themselves that the infinite regress of Repetition (whether in the sense of externally generated serial writing or of mass reproduction) can be escaped.

In the scant seven years since *Noise* was published, extraordinary evidence of such tendencies in music has emerged. It was in the mid-1970s that New Wave burst on the scene in England, with precisely the motivation suggested by Attali at his most optimistic and with the mixed results he also realistically anticipated. Many of the original groups began as garage bands formed by people *not* educated as musicians who intended to defy noisily the slickly marketed "nonsense" of commercial rock. The music is often aggressively simple syntactically, but at its best it conveys most effectively the raw energy of its social and musical protest. It bristles with genuine sonic noise (most of it maintains a decibel level physically painful to the uninitiated), and its style incorporates other features that qualify as cultural noise: the bizarre visual appearance of many of its proponents, texts with express political content, and deliberate inclusion of blacks and of women (not as the traditional "dumb chicks" singing to attract the libidinous attention of the audience, but—taboo of taboos—as competent musicians *playing instruments,* even drums).

The grass-roots ideology of the New Wave movement has been hard to sustain, as the market has continually sought to acquire its products for mass reproduction. Even among the disenfranchised, the values of capitalism are strong, and many groups have become absorbed by the recording industry. The realization that much of their most ardent protest was being consumed as "style" caused a few groups, such as the Sex Pistols, to disband shortly after they achieved fame. But while there exists a powerful tendency for industry to contain the noise of these groups by packaging it, converting it into style-commodity, the strength of the movements is manifested by the seeming spontaneous generation of ever more local groups. The burgeoning of Composition, still somewhat theoretical in Attali's statement of 1977, has been actualized and is proving quite resilient.

The same seven-year period has witnessed a major shift in "serious" music, away from serialism and private-language music toward music that strives once again to communicate. Whether performance art, minimalism, or neo-tonality, the new styles challenge the ideology of the rigorous, autonomous, elitist music produced in universities for seminars. They call into question the institutions of academic training and taxonomies, of orchestras and opera houses, of recording and funding networks.

Many of the principal figures in these new styles come from groups traditionally marginalized, who are defined by the mainstream as noise anyway, and who thus have been in particularly good positions to observe the oppressive nature of the reigning order. Women, for instance, are not only strongly represented in these new modes of Composition—they are frequently leaders, which has never before been the case in Western "art" music. Instead of submitting their voices to institutionalized definitions of permissible order, composers such as Laurie Anderson and Joan La Barbara celebrate their status as outsiders by highlighting what counts in many official circles as noise. Some individuals composing new kinds of music were originally associated with other media (David Hykes with film, for instance) or have found their most responsive audiences among dancers and visual artists (Philip Glass). All are people who managed not to be silenced by the institutional framework, who are dedicated to injecting back into music the noise of the body, of the visual, of emotions, and of gender.

For the most part, this music is far more vital than the music of Repetition, which has deliberately and systematically drained itself of energy. Many practitioners of Composition fight the tendency toward objectification by making live, multimedia performance a necessary component of the work. Others (such as Pauline Oliveros) explore the possibility of breaking down the barrier between producer and consumer by designing instructions for participatory events. Collaborative efforts (combining music, drama, dance, video) are prominent in these movements. The traditional taxonomic distinction between high and popular culture becomes irrelevant in the eclectic blends characteristic of this new music, and indeed many of these new composers are as often as not classified as New Wave and perform in dance clubs. A new breed of music critic (such as John Rockwell and Gregory Sandow) has begun to articulate the way the world looks (and sounds) without the distortion of that distinction.

Composition, as Attali defines it, is coming increasingly to the fore, displacing the musical procedures and the networks of Repetition. That these new movements signal not simply a change in musical taste but also of social climate seems extremely plausible, though how exactly the change will be manifested in other areas of culture remains to be seen. At the very least the new movements seem to herald a society in which individuals and small groups dare to reclaim the right to develop their own procedures, their own networks. *Noise*, by accounting theoretically for these new ways of articulating possible worlds through sound and by demonstrating the crucial role music plays in the transformation of societies, encourages and legitimates these efforts.

Notes

Notes

Foreword

1. Max Weber, "Value-judgements in Social Science," in *Max Weber: Selections in Translation*, edited by W. G. Runciman (Cambridge: Cambridge University Press, 1978), p. 95. See also Weber's lengthier discussion in *The Rational and Social Foundations of Music*, trans. Martindale, Riedel, and Neuwirth (Carbondale: Southern Illinois University Press, 1958).

2. "Value-judgements," p. 96.

3. Ibid., p. 96.

4. Theodor W. Adorno. *Introduction to the Sociology of Music*, trans. E. B. Ashton (New York: Seabury, 1976), p. 62.

5. Ibid., p. 63.

6. Ibid., p. 209. On the question of the autonomous status of art, see also in this series, Peter Bürger, *Theory of the Avant-Garde*, introduction by Jochen Schulte-Sasse, trans. Michael Shaw, vol. 4 (Minneapolis: University of Minnesota Press, 1984).

7. Oswald Spengler, *The Decline of the West*, vol. 1 (New York: Knopf, 1939), p. 47.

8. Theodor W. Adorno, *Philosophy of Modern Music*, trans. Mitchell and Blomster (New York: Seabury, 1973), p. 66.

9. Jacques Attali, *Les Trois mondes* (Paris: Fayard, 1981), p. 270

Chapter 1. Listening

1. Friedrich Nietzsche, *The Birth of Tragedy*, trans. Walter Kaufman (New York, Vintage, 1967), p. 63.

2. Nietzsche, *Birth of Tragedy*, p. 61.

3. "Whether we inquire into the origin of the arts or observe the first criers, we find that everything in its principle is related to the means of subsistence." Jean-Jacques Rousseau, *Essai sur l'inéqalité*.

4. Gottfried Wilhelm Leibnitz, "Drôle de pensée touchant une nouvelle sorte de représen-

tation," ed. Yves Belaval, *La Nouvelle Revue Francaise* 70 (1958): 754-68. Quoted in Michel Serres, "Don Juan ou le Palais des Merveilles," *Les Etudes Philosophiques* 3 (1966): 389.

5. [A reference to Rabelais, *Gargantua and Pantagruel*, b. 4, chap. 54. TR.]

6. [Andrei A. Zhdanov, "Music," address to a conference of leading musicians called by the Central Committee of the Communist party, 1947. Published in *Essays on Literature, Philosophy and Music* (New York: International Publishers, 1950), pp. 81, 95-96 (emphasis added). TR.]

7. Fritz Stege, *La situation actuelle de la musique allemande* (1938).

8. Michel Serres, *Esthétique sur Carpaccio* (Paris: Hermann, 1975).

9. Ibid.

10. Jean Baudrillard, *L'échange symbolique et la mort* (Paris: Gallimard, 1976), p. 116.

11. Nietzsche, *Birth of Tragedy*, p. 119.

12. See Dominique Zahan, *La dialectique du verbe chez les "Bambaras"* (Paris: Mouton, 1963).

13. Serres, *Esthétique*.

14. René Girard, *Violence and the Sacred*, trans. Patrick Gregory (Baltimore: Johns Hopkins University Press, 1977).

15. Serres, *Esthétique*.

16. Richard Wagner, "The Revolution," in *Prose Works*, trans. William Ashton Ellis, vol. 8 (New York: Broude Bros., 1966), p. 237.

17. Marina Scriabine, *Le langage musical* (Paris: Minuit, 1963).

18. Montesquieu, *The Spirit of the Laws*, trans. Thomas Nugent (New York: Hafner Publishing Co., 1965).

19. Jacques-Gabriel Prod'homme, ed., *Ecrits des musiciens* (Paris: Mercure de France, 1912), pp. 351-60.

20. Hans David and Arthur Mendel, eds., *The Bach Reader* (New York: Norton, 1945), pp. 51-52.

21. [Aristotle, *Politics*, 1339a-1342b. TR.]

22. Roland Barthes, "Le grain de la voix," *Musique en Jeu* 9 (Nov. 1972): 51-63.

23. Carlos Castaneda, *The Teachings of Don Juan* (Berkeley and Los Angeles: University of California Press, 1971), pp. 57 and 60.

Chapter 2. Sacrificing

24. We are discussing Brueghel's painting, *Carnival's Quarrel with Lent*, a reproduction of which is found on the cover.

25. Luigi Russolo, "The Art of Noises" (1913), in *Futurist Manifestos*, ed. Umbro Apollonio (New York: Viking, 1973), pp. 74-75.

26. Once again, *Carnival's Quarrel with Lent* displays a fascinating symbolism of the chair, which is in turn an attribute of power and tool of repentance.

27. At this point, this chapter becomes more theoretical and may be skipped without inconvenience on the first reading.

28. Claude Lévi-Strauss, *The Naked Man*, trans. John and Doreen Weightman (London: Jonathan Cape, 1981), pp. 659 and 647.

29. Jean-Jacques Rousseau, *Essai sur l'origine des langues*, in *Oeuvres Complètes*, Bibliothèque de la Pléiade, vol. 2 (Paris: Gallimard, 1962).

30. Karl Marx, *Capital*, trans. Ben Fowles, vol. 1 (New York: Vintage, 1977), p. 284 (bk. 1, chap. 7).

31. Gottfried Wilhelm von Leibnitz, "On the Radical Origination of Things" (1697), in *Philosophical Papers and Letters*, ed. and trans. Leroy Loemker, 2d ed. (Dordrecht, Holland: D. Reidel, 1969), p. 490.

32. Charlie Gillett, *The Sounds of the City* (New York: Outerbridge and Dientstrey, 1970), p. 300. The reference is to Colin Fletcher, *The New Society and the Pop Process* (London, 1970).

33. Ssu-ma Ch'ien, *Les mémoires historiques de Se-ma Ts'ien*, trans. Edouard Chavannes, vol. 3 (Paris: Leroux, 1895-1905), pp. 239-40 (chap. 24).

34. *Book of Rites*, trans. James Legge, vol. 2 (1882; reprint, New Hyde Park, NY: University Books, 1967), p. 276 (chap. 25, par. 15). Emphasis added. [Translation modified to agree with the French version of S. Couvreur, *Li Ki, ou Mémoires sur les bienséances et les cérémonies*, vol. 2 (Ho Kien Fu: Imprimerie de las Mission Catholique, 1893), p. 897 TR.]

35. A. R. Radcliffe-Brown, *The Andaman Islanders* (New York: The Free Press, 1964), p. 132. [Translation modified to agree with Attali's French translation. TR.]

36. Henri Atlan, *Organisation en niveau hiérarchique et information dans les systèmes vivants* (Paris: ENSTA, 1975).

37. Plato, *The Republic*, trans. Paul Shores, Loeb Classical Library (New York: Putnam's, 1930), pp. 333-34 (bk. 4, sec. 424).

38. Marx, *Capital*, sec. 4, chap. 15.

39. Girard, *Violence*, (note 14), p. 27.

40. Karl Marx, *Theories of Surplus Value*, trans. Emile Burns, vol. 1 (Moscow: Foreign Languages Publishing House, 1963).

41. Ibid., pp. 394-95 (addendum 12, F, 1328).

42. Karl Marx, "Travail productif et travail improductif," *Matériaux pour l'économie*, chap. 2, in *Oeuvres complètes*, Edition de la Pléiade, vol. 2 (Paris: Gallimard, 1968), p. 393.

43. [In French, *représentation* covers the English "performance" as well as "representation" in the philosophical sense. In order to preserve this connection (and the distinction between *représentation* and *exécution*), "representation" has been retained for both connotations, in spite of the occasional awkwardness this produces in English. For the author's definition of the terms *representation* and *repetition* (which also has an additional connotation in French: "rehearsal"), see p. 41 below. TR.]

44. There is an extensive literature on this theme. Many economists have even thought it possible to restrict the study of the economics of art to it:

David Berger and Richard Peterson, "Entrepreneurship in Organizations: Evidence from the Popular Music Industry," *Administrative Science Quarterly* 16, no.1 (1971): 97-108.

David Lacy, "The Economics of Publishing, or, Adam Smith and Literature," *Daedalus* 92, no. 1 (1963): 42-62.

W. J. Baumol, and W. G. Bowen, *Performing Arts: The Economic Dilemma* (Cambridge, Mass.: MIT Press, 1968).

David Berger and Richard Peterson, "Three Eras in the Manufacture of Popular Lyrics," in Denisoff and Richardson, *The Sounds of Social Change* (see below).

Especially, Serge R. Denisoff, "Protest Movements: Class Consciousness and Propaganda Songs," *Sociological Quarterly* 9, no. 2 (1968): 228-47; Denisoff, "Folk Music and the American Left," *British Journal of Sociology* 20, no. 4 (1968): 427-42; Denisoff and Peterson, eds., *The Sounds of Social Change* (Chicago: Rand McNally, 1972).

D. J. Hatch and D. R. Watson, "Hearing the Blues," *Acta Sociologica* 17, no. 2 (1974): 162-78. From an ethnological point of view: A. Lomax, "Song Structure and Social Structure," in *The Sociology of Art and Literature: A Reader*, ed. Milton Albrecht, James Barnet, and Mason Griff (London: Duckworth, 1970), pp. 55-71.

Max Weber, *Rational and Social Foundations of Music*, trans. Don Martindale, Johannes Riedel, and Gertrude Neuwirth (Carbondale, Ill.: Southern Illinois University Press, 1958). See also the preface to *The Protestant Ethic and the Spirit of Capitalism*, trans. Talcott Parsons (New York: Scribner's, 1958).

A. P. Merriam, "Music in American Culture," *American Anthropologist* 57, no. 6 (1956): 1173-81; K. P. Etzkorn, "On Music, Social Structure and Sociology," *International Review of the Sociology of Music*, vol. 5 (1974).

45. Karl Marx, "Introduction to a Critique of Political Economy," in *The German Ideology*, ed. C. J. Arthur (New York: International Publishers, 1970), p. 149.

46. Theodor Adorno, *The Philosophy of the New Music*, trans. Anne Mitchell and Wesley Blomstel (New York: Seabury Press, 1973).

47. Theodor Adorno, *Dissonanzen* (Götingen: Vandenhoeck and Ruprecht, 1963).

48. Gilles Deleuze and Félix Guattari, "Rhizome," trans. Paul Foss and Paul Patton, in *Ideology and Consciousnes* 8 (Spring 1981): 63.

49. Castaneda, *Teachings*, p. 88.

Chapter 3. Representing

50. David and Mendel, *Bach Reader* (note 20), pp. 49-50.

51. Karl Geiringer, *Haydn: A Creative Life in Music* (New York: Norton, 1946), p. 52.

52. Dedication to *Persée: Tragédie mise en musique par M. de Lully, Escuyer, Conseiller Secrétaire du Roy, et Surintendant de la Musique de Sa Majesté*, in Prod'homme, *Ecrits des musiciens* (note 19), p. 209.

53. *Le bourgeois gentilhomme*, act I, scene ii.

54. Jean-François Marmontel, "Epître dédicatoire," *Encyclopédie*, vol. 5 (1751-65), p. 822.

55. Julien Tiersot, *Lettres de musiciens écrites en francais du XVe au XVIIIe siècle*, vol. 1 (Turin: Boca, 1924), p. 99.

56. Albert Lavignac and Lionel de la Laurencie, eds. *Encyclopédie de la musique et dictionnaire du conservatoire*, vol. 1 (Paris: Librairie Delagrave, 1913-34), p. 1565.

57. Letter to his father, May 1, 1778, in *The Letters of Mozart and His Family*, ed. E. Anderson, 2d ed., vol. 2 (New York: St. Martin's Press, 1966), p. 531.

58. Stéphanie de Genlis, *Les soupers de la maréchale de Luxembourg*, 2d ed., vol. 3 (Paris, 1828), p. 44. Cited in Judith Tick, "Musician and Mécène: Some Observations on Patronage in Late Eighteenth Century France," *International Review of the Aesthetics and Sociology of Music* 4, no. 2 (1973): 247.

59. Gertrude Norman and Miriam Shrifte, eds. *Letters of the Composers* (New York: Knopf, 1946), pp. 22-23.

60. *Pétition pour la création d'un Institut national de Musique*, read from the stand by Bernard Sarette to the National Convention on 18 Brumaire, year II (manuscript, Archives Nationales F1007, no. 1279).

61. *Règlement de l'Institut national de Musique* (extract), 1793.

62. "Argumentaire de Gossec et autres" (Archives Nationales, A XVIIIe, 384).

63. Ibid.

64. *Règlement de l'Institut national de Musique*.

65. Ibid.

66. "Argumentaire de Gossec et autres."

67. Ibid.

68. For a long time, the piracy of musical works went unpunished. The Tribunal of the Seine district, on May 30, 1827, looked upon it with indulgence: it found no evidence of piracy "in two or three contredanses based on the score of Rossini's *Siège de Corinthe*, only simple plagiarism, because the author transformed the rhythm to adapt it into a quadrille, which the author of the opera did not himself bring to realization." Rossini had not been wronged "since he published neither fantasia nor quadrille based upon that piece."

69. An outline of this idea can be found in Michel Foucault, *The Order of Things* (New York: Pantheon, 1970), p. 190.

70. Plato, *Laws*, trans. R. G. Bury, Loeb Classical Library, vol. 2 (Cambridge, Mass.: Harvard University Press, 1952), p. 129 (bk. 2, 665a).

71. Ssu-ma Ch'ien (2d century B.C.), *Les mémoires historiques de Se-ma Ts'ien*, trans. Edouard Chavannes, vol. 3 (Paris: Leroux, 1895), p. 252 (chap. 24).

72. Jan Pieterszoon Sweelinck, *Cinquante psaumes de David*, dedication to the burgomasters and aldermen of the city of Amsterdam, in Prod'homme, *Ecrits de musiciens* (note 19), p. 102.

73. *Essai sur l'origine des langues* (note 29), chap. 20: "I say that any language that cannot be understood by the people assembled is a servile language; it is impossible for a people to speak such a language and remain free."

74. Shakespeare, *Troilus and Cressida*, act I, scene iii, lines 109-10.

75. David and Mendel, *Bach Reader* (note 20), pp. 118-19.

76. From Charles Burney, *The Present State of Music in France and Italy* (1771), quoted in Oliver Strunk, *Source Readings in Music History* (New York: Norton, 1950), pp. 689 and 693.

77. Scriabine, *Le langage musical* (note 17).

78. Abbé Jean Antoine Dubois, *Description of the Character, Manners, and Customs of the People of India: and of their Institutions, Religious and Civil*, vol. 2 (Philadelphia: McCarey and Son, 1818), pp. 162-63.

79. See Kinsky's catalog.

80. Hector Berlioz, *Treatise on Instrumentation, including Berlioz' Essay on Conducting*, ed. Richard Straus, trans. Theodore Front (New York: Edwin Kalmus, 1948), p. 420. [Translation modified. TR.]

81. Ibid., p. 416. [Translation modified. TR.]

82. Lavignac and Laurencie, *Encyclopédie de la musique* (note 56), p. 2133.

83. Frederick Goldbeck, *The Perfect Conductor* (London: Dennis Dobson, 1960), p. 113.

84. Letter of Muzio Clemente, London music publisher, to his associate F. W. Collard, dated April 22, 1807, from Vienna, in Norman and Shrifte, *Letters of Composers* (note 59), pp. 65-66.

85. Letter to Puchberg, ibid., p. 79.

86. Jean and Brigitte Massin, *Mozart* (Paris: Fayard, 1971), pp. 577-81.

87. Norman and Shrifte, *Letters of Composers*, pp. 147-48.

88. Richard Wagner, "Du métier du virtuose" (article in the *Revue et Gazette Musicale*, October 18, 1840), reproduced in *Prose Works* (note 16), vol. 7, pp. 127-28.

89. Quoted in Mina Curtis, *Bizet et son temps* (Paris: La Patine, 1961).

90. Document cited by Patrice Coirault in *Formation de nos chansons folkloriques* (Bibliothèque Nationale, manuscript 22116).

91. *Revue musicale* 37 (1835).

92. Georges Coulonges, *La chanson en son temps: De Béranger au juke-box* (Paris: Editeurs Français Réunis, 1969).

93. Ibid.

94. Ibid.

95. "Finding that the law does not measure its protection by the length of the production, and its actions are universal; that these actions have as their object the goal of preserving a man's right to his thought and to compensation for labors that honor the intelligence . . . that this principle merits all the more respect since a possession that a judge could recognize or deny at whim and at a moment's fancy is no possession at all, and that additionally the power conferred upon the courts to take as the criterion of its decision the stature of the usurped work would lead to the most glaring injustices." *Receuil Dalloz*, vol. 53, sec. 2, p. 13.

96. *La France Musicale* 10 (March 10, 1850).

97. Law of March 11, 1857.

98. For mechanical reproduction, the publisher generally receives 50 percent and the intellectual creators 50 percent. This indicates, as we will see later on, that the balance of power between the various parties changed with the sudden emergence of mechanical means of reproduction.

99. See Dominique Jameux, *Musique en jeu* 16 (Nov. 1974): p. 55.

100. Stefan Zweig, *The World of Yesterday* (New York: Viking, 1943), p. 22.

101. See a research note by Y. Stourdzé (Paris: IRIS, 1976).

Chapter 4. Repeating

102. See Georges Dumézil, *Mythes et épopées* (Paris: Gallimard, 1968).

103. Gilbert Arthur Briggs, *Reproduction sonore à haute fidélité* (Paris: Société des Editions de Radio, 1958), p. 10.

104. *Langue musicale universelle, inventée par F. Sudre, également inventeur de la téléphonie musicale*, with *Le vocabulaire de la lanque musicale* (Tours, 1866). Cited in Louis Figuier, *Merveilles de la Science* (Paris: Furne, Jouvette, 1867).

105. *La Nature*, April 25, 1891, p. 253.

106. Thomas Alva Edison, in *The Phonogram* 1 (New York, 1891-93): 1-3.

107. Cited by Th. Reinach, deputy, report no. 3156 to the Chamber of Deputies, March 1, 1910.

108. Figures cited by Th. Reinach, deputy, in report no. 3156 to the Chamber of Deputies (March 1, 1910), on a bill relating to the protection of authors in the reproduction of works of art.

109. *Gazette des Tribunaux* 2, no. 3 (1901).

110. *Le Droit*, March 5, 1905.

111. On March 22, 1927, M. Privat, manager of the Eiffel Tower, was ordered by the civil court of the Seine district to pay 3,000 francs in damages for having broadcast without authorization pieces of music listed with SACEM. On appeal, the First Chamber of the court of Paris tripled the indemnity.

112. Extract from the *Revue Musicale*, 1920; cited in Philippe Carles and Jean-Louis Comolli, *Free Jazz et Black Power* (Paris: Champ Libre, 1974), pp. 65-66.

113. Interview with Denis Constant, *Jazz Magazine* (Paris) 203 (February 1973).

114. Cited in Philippe Daufouy and Jean-Pierre Sarton, *Pop-music/Rock* (Paris: Champ Libre, 1972).

115. Cited in Rolf Liberman, *Actes et entr'actes* (Paris: Stock, 1976).

116. [In French, the publishing term *édition* covers the reproduction and distribution of any intellectual work, whether literary or musical, printed or recorded. TR.]

117. See the report of J.-L. Tournier, general director of SACEM, to the Thirtieth Congress of CISAC (Confédération Internationale des Sociétés d'Auteurs et Compositeurs), Sept. 27, 1976.

118. John Cage, *Silence* (Middletown, Conn.: Wesleyan University Press, 1961), p. 76.

119. Claude Lévi-Strauss, *The Raw and the Cooked*, trans. John and Doreen Weightman (New York: Harper and Row, 1969), p. 22.

120. Earle Brown, interview, *Darmstädter Beiträge zur neuen Musik* 10 (1966): 58.

121. Iannis Xenakis, *Musique et architecture* (Paris: Casterman, 1971), p. 32.

122. Karlheinz Stockhausen, *VH* 101 (Winter 1970-71).

123. Gian Franco Sanguinetti Censor, *Véridique rapport sur les dernières chances du capitalisme en Italie* (Paris: Champ Libre, 1976).

124. Ibid.

125. Lucien Sfez, *Critique de la décision* (Paris, A. Colin, 1976), pp. 160 and 330.

Chapter 5. Composing

126. Roland Barthes, *L'Arc* 40: p. 17.

127. Apollonio, *Futurist Manifestos* (note 25), p. 85.

128. Cage, *Silence*, p. 10.

129. "Freedom" in Swahili.

130. Carles and Comolli, *Free Jazz* (note 112).

131. Interview with Billy Harper, *Jazz Magazine* (Paris) 215 (September 1973): 43-44.

132. Interview with Clifford Thornton, *Jazz Magazine* (Paris) 208 (February 1973): 14.

133. Extract from a speech by Malcolm X to the Organization of African Unity, June 28, 1964, in Frank Kofsky, *Black Nationalism and the Revolution in Music* (New York: Pathfinder Press, 1970, pp. 65-66.

134. Interview with Archie Schepp, *Jazz Magazine* (Paris) 243 (April 1976): 16.

135. Marx, *Capital*, vol. 1 (note 30), 1, chap. 7.

136. Sigmund Freud, *The Moses of Michelangelo*, standard edition, trans. James Strachey, vol. 13 (London: Hogarth, 1958), p. 211.

137. See B. Lassus, "Les habitants paysagistes" (unpublished manuscript).

138. Claude Lévi-Strauss, *The Raw and the Cooked* (note 119), p. 18.

Index

Index

AACM, 138
Academy of Music (France), 56
Admission charges, 32, 50, 57, 58, 79
Adorno, Theodor, ix, x–xi, 11, 30, 43, 104,
    115, 153; *Negative Dialectics*, xi
Advertising. *See* Marketing
Aleatory music, 34, 114–15, 120
Alienation, 134–35, 138, 141, 142
Althusser, Louis, vii
Ambrosian liturgy, 13
American Graphophone Company, 95
American Society of Composers, Authors,
    and Publishers (ASCAP), 105
Anderson, Laurie, 157
Android, 85
Annales, vii
Antheil, George, 136
Apprenticeship, 63
Aristotle, 18
Arpeggione, 35
Arras, 52
Arte Dei Rumori (Russolo), 10
ASCAP, 105
Association for the Advancement of Creative
    Musicians (AACM), 138
Atonal music, 119
Attali, Jacques, vii, xi–xiv, 149–58; as

economist, xii; *Les Trois Mondes*, xii;
    utopianism of, xiii, xiv
Auerbach, Erich, vii
Automobile: and noise, 123–24
Autosurveillance, xiii

Bach, Johann Sebastian, 5, 17–18, 19, 28,
    35, 47–48, 63, 68, 115, 131, 146, 151
Beleine, M. (cabaret owner), 75
Ballard, Robert, 53, 54
Ballet mécanique (Antheil), 136
Banalization: of music, 109; of objects, 120
Bannister, John, 50
Barrel organ, 73
Barthes, Roland, 18, 135
Bataille d'Austerlitz (Gérard), 197
Baton, 67
Baudrillard, Jean, xii, 83
Beatles, 110, 119, 125
Beaver, Paul, 139
Beethoven, Ludwig van, ix, xi, 28, 32, 35,
    50, 66, 69–70, 92, 100, 152
Belief: and music, 19, 46, 57, 59, 61, 154,
    155
Bell, Daniel, xii
Berg, Alban, 28
Berio, Luciano, 141

Berliner, Emile, 95
Berlioz, Hector, 11–12, 66–67
Bible: Old Testament, 27; Tables of the Law, 87
Billboard, 107
Bird organ (*serinette*), 73
Bizet, Georges, 71
Black Muslims, 139
Blaue Reiter, Der, 81
Bley, Carla, 138
Body, 32, 142–43, 144
Bollecker, L., 99
Boukay, Maurice (M. Couyba), 76
Boulez, Pierre, 39, 113, 145–46
Bourgeoisie, 43, 47, 50, 55, 56–57, 58, 60, 61–62, 69, 75, 82, 119, 135. *See also* Class
Bourget, E., 77
Bowie, David, 119
Brahms, Johannes, 151
Braut (Ries), 71
Brueghel, Pieter, 21–24, 130, 148. *See also Carnival's Quarrel with Lent*
Burney, Charles, 63

Cabaret, 72–75, 76
Caccini, Giulio, 155
"Cadet Roussel" (popular air), 73
Café concert, 72, 74, 75–77
Café Concert des Ambassadeurs, 77
Café des Aveugles, 75
Cage, John, 18, 112, 114, 116, 136–37
Caillois, Roger, 137
Call for the Emancipation of Dissonance, A (Der Blaue Reiter), 81
Cannabich, Christian, 43
Capitalism, 42, 44, 131; contemporary, xii, 78, 96; and music, 12, 16, 24, 36–42, 51, 81, 83–84, 88, 102-3, 107, 121, 138–39, 154, 156, 157; self-destruction of, 5; and serial decomposition, 65
Carnival's Quarrel with Lent, 21–24, 45, 81, 106, 119, 120, 121, 124, 130, 141, 142, 147, 148, 162 n26
Caruso, Enrico, 98
Cash Box, 107
Castaneda, Carlos, 20, 44, 147
Cavatina, 74
Caveau, Le (cabaret), 75, 76

Censor, Gian Franco Sanguinetti, 121
Censorship, 8, 105, 109, 131, 138
Centralization, 55–57, 129
Cesti, Antonio, 155
Chansonnier, 75, 76
Charlemagne, 14
Charles, D., 141
Chat Noir (cabaret), 76
Chaussettes noires (singing group), 106
Chavez, Carlos, 136
Cherubini, Carlo, 6, 56, 70
Chevalier, Maurice, 76
China, 13, 29, 60, 65
Choiseul, count of, 50
Chopin, Frédéric, 70–71
Clapton, Eric, 109
Claque, 77
Clarinet, 73
Class: and music, 14, 16, 43, 46, 62, 67, 69, 75, 76, 119, 140, 155. *See also* Bourgeoisie
Classical music: and repetition, 85. *See also* "Serious" music
Clé du Caveau, La (journal), 72
Clément, J.-B., 76
Codes: dynamics of, 31–34; liquidation of, 33–36, 42, 44, 45, 122, 137, 142; music as stockpiling of, 30; mythological, 28; sacrificial, 24–27, 31; simultaneity of, 43–44; and time, 147
Combinatorics, 64–65, 81–86, 89
Commission for the Renovation and Development of Musical Studies (France), 64, 94
Commodity. *See* Music, as commodity
Commodity-value, 37
Communication, 9, 32, 36, 121, 122, 134, 142, 143, 157
Compagnie Française du Gramophone, 95
Compagnie des Gramophones, 98
Competition, 51, 68, 128
Composers: and copyright, 52, 53–54; as molders, 37, 40; and productive labor, 39–40; in repetition, 115; in representation, 39; and royalties, 39
Composition, 20, 32, 33, 36, 45, 132, 150, 156–58; New York schools of, 153
Concert des Amateurs (Gossec), 51
Concert hall, 47, 50, 55, 57, 73, 117–18, 120
Concert spirituel (Philidor), 51

Concerto, 67
Condorcet, Antoine de Caritat, Marquis de, 64
Conductor. *See* Orchestra leader
Confinement: of music, 32, 63, 74–75, 76; of youth, 109
Confucius, 29–30
Conservatories, 63–64
Conservatorio di Sant' Onofrio, 63
Conservatory of Music (France), 56
Construction (Cage), 116
Copyists, 52, 53
Copyright: in representation, 52–54, 68, 72, 74, 77, 79, 84, 165 n95; in repetition, 96–101, 166 n111
Coriolan (Beethoven), 69
Cosi Fan Tutte (Mozart), 143
Council of Avignon, 22
Council of Bayeux, 22
Council of Paris, 22
Counterpoint, 9, 151
Court music, 14, 17
Couyba, M., 76
Crisis: in networks of music, 127; of proliferation, 45, 128–32, 135; in repetition, 130–31
Cros, Charles, 90–91

Dance: popular, 80, 118, 119
Danican, François-André (Philidor), 51
Danzi, Franz, 43
Death: in composition, 143; monopolization of, 30; music and threat of, 120; noise as cause of, 27; stockpiling of, 125
Debussy, Claude, 28
Decentralization, 129, 146
Dedication, 17, 48, 49
Demand: production of, 42, 101, 103, 106–9, 127, 128–29, 130, 131, 146
Democracy, music in, 8–9, 156
Demoiselles d'Avignon, Les (Picasso), 81
Depersonalization, 114
Derrida, Jacques, 25
Descartes, René, 151
Difference: loss of, and music, 5, 106, 119; music as bulwark against, 8; and noise, 22; and order, 62; recreated, 26, 28, 45, 142, 145, 147; and socialization through identity, 110, 121

al-Din Runir, 27
Dissonance, 27, 29, 35, 43, 61, 62, 83, 130, 155
Distribution: of music, 8, 31, 35, 46, 109, 122
Dixon, Bill, 138
Donay, Maurice, 76
Dranem (comic), 76
Drugs, 20
Dylan, Bob, 6

Economics: and music, xi, 5, 10, 39. *See also* Exchange, in composition; Harmony, and exchange; Money; Music, as commodity; Music, as exchange
Edison, Thomas Alva, 90, 91–92, 93–94, 95, 103
Electroforming, 94
Electronic music, 9
Elitism, 113, 115–16, 117, 118, 119
Emerson, Keith, 109
Encyclopédie de la musique (Lavignac and Laurencie), 67
Engels, Friedrich, xi
Equal temperament, 61
Esterházy, Johann Nepomuk, Count, 48
Euler, Leonhard, 33
Exchange, 24, 57–58, 68, 143, 145, 147; in composition, 143, 145; destroys sacrifice, 134, 141; destroys usage, 42, 81, 84, 88, 107–8; and harmony, 59–65; of signs, 53. *See also* Music, as exchange
Exchange-time, 58, 101, 124, 125, 129, 132, 135
Exchange-value, 37, 59, 72, 77, 84, 97, 99, 101, 107, 141, 142
Expression: music as, ix, x

Family: death of, 110
"Fanchon" (popular air), 73
Faust (Spohr), 71
Fête Ininterrompue, La, 51
Fêtes de Thalie, 51
Fétis, E., 73–74
Fifth Symphony (Beethoven), 92
Filitz (composer), 43
Flageolot, 73
Flagstad, Kirsten, 106
Fletcher, Colin, 27

Forgetting: music as, 19
Foss, Lukas, 116
Foucault, Michel, vii
Fourier, Jean Baptiste Joseph, Baron de, 35, 65
Fourth Symphony (Beethoven), 69
France Musicale, La (journal), 78
François, Sudre, 92
Frankfurt School, ix, x
Franklin, Benjamin, 35
Frederick II (king of Prussia), 16
Free jazz. *See* Jazz, free
Freedom and Unity (Thornton), 139
Freie Satz, Der (Schenker), 151
French Revolution, 54–56
Freud, Sigmund, 6, 143

Gérard, François, 97
Gesualdo, Carlo, 18
Gil Blas Illustré (magazine), 74
Gille, C. (*goguette* owner), 75
Girard, René, 10
Glass harmonica, 35
Glass, Philip, 114–15, 157
Goguette, 75
Goodman, Benny, 104
Gossec, François-Joseph, 51, 55, 56
Gottschalk, Louis Moreau, 71
Gramophone, 94, 101
Grateful Dead, 105
Greece, x, 13, 29, 60, 65, 150, 155
Gregorian chant, 14, 15
Grétry, André-Ernest-Modeste, 49, 60
Guilds, 15–16
Guitar, 74; electric, 35

Habermas, Jürgen, 91
Half-tone, 10
Handel, George, 50–51
Harmony, viii, 9, 10, 19, 21, 27, 35, 46, 57, 81, 83, 89, 130; anti-, 83; and exchange, 59–65; Rameau's theory of, 150–51
Harp, 73
Harpsichord: banned, 56
Haydn, Franz Joseph, 43, 48, 66, 111
Hegel, Georg Wilhelm Frederick, vii, x
Hegelianism, neo-, 151–52
Helmholtz, Hermann von, 35
Hendrix, Jimi, 6, 35, 105, 109, 125, 137
Henrion, P., 77

Hindemith, Paul, x
Historicism, vii, x, xi, xii
Hit parade, 4, 6, 88, 106–9
Hit songs, 76, 108; ranking of, 106–8
Hitler, Adolf, 87
Holzbauer, Ignaz, 43
Homosexuality, 13
Honegger, Arthur, 136
Howe, Steve, 109
HP (Chavez), 136
Hykes, David, 157

Identification, 45, 67, 110, 118–19, 120, 121, 130
Ideology: and music, ix, xi, 19, 31, 55, 112, 116, 153, 156
Improvisation, 114, 140, 145–46, 155
L'Inconnue, 51
Incontri (Nono), 116, 133
Industrial revolution, 144
Information theory, 26–27, 33
Ingres, Jean Auguste Dominique, 97
Instruments, musical: invention of, vii, 35, 140–41, 144, 145; mechanical, 73, 97. *See also* Arpeggione; Barrel organ; Flageolot; Glass harmonica; Guitar; Harp; Harpsichord; Oboe; Piano; Violin
Interpreters. *See* Performers
Iron Foundry, The (Mossolov), 136

Jazz, 8, 27, 103–5, 109; free, 137, 138–40
Jazz Composers' Guild, 138
Jazz Composers' Orchestra (JCOA), 138
JCOA, 138
Jefferson Airplane, 105
Jessonda (Spohr), 71
Jongleur, 14, 15, 17, 31, 39, 63, 141
Joplin, Janis, 6, 106, 125
Jukebox, 84, 93, 95, 104

Kagel, Mauricio, 116
Kenton, Stan, 104
Keynesian economics, 130, 131
Klein, Yves, 137
Knowledge: and music, 4, 9, 18–20, 133, 150

La Barbara, Joan, 157
Labor, productive, 37–41
Labor-value, 58–59, 89
Lamentations, 23, 133

Language, 25; musical (universal), 92, 116
Late Capitalism (Mandel), xii
Laurencie, Lionel de la, 67
Lavignac, Albert, 67
"Learned" music. *See* "Serious" music
Lebkowitz, Prince, 50
Leibnitz, Gottfried, 27; "Palace of Marvels," 6–7
Le Roy, Adrien, 53
Lesueur, Jean-François, 70
Lévi-Strauss, Claude, vii, 28–29, 146
L'Hôpital, Charles, 64
Lice Chansonnier, La (*goguette*), 75
Light music. *See* Popular music
Lip syncing, 118
Liszt, Franz, 68–69, 70, 71
Liturgical music, 10, 14
Loudspeaker, 87
Louis XI (king of France), 52
Louis XIV (king of France), 48, 155
Louis XV (king of France), 54
Luigini, M. (conductor), 94
Lully, Jean-Baptiste, 48–49, 51, 54
Lyotard, Jean-François, xii, 35

Mahler, Gustav, 81, 82, 152
Mandel, Ernest, xii
Mantler, Mike, 138
Marketing (publicity, advertising), 85, 96–97, 106–9, 129
Markov, Andrey Andreyvich, 35
Marmontel, Jean-François, 49
Marx, Karl, xiii, 6, 38, 39, 42–43, 58
Marxism, vii, xi, xii-xiii, 37–39, 43, 58, 62, 89, 110, 128
Mass media, xii, 108, 131, 137
Mass production, 32, 68, 84, 87, 89, 102–3, 105, 106, 121, 128
Match (Kagel), 116
Mathematics, 35, 65, 113, 150, 151
Mathilde, Princess, 71
McDonald, "Country" Joe, 105
Meaning: and music, 25, 33, 44–45, 83, 102–3, 112–17, 122, 142, 147, 152
Media. *See* Mass media
Méhul, Etienne-Nicolas, 56
Melba, Nellie, 98
Ménagerie, La (*goguette*), 75
Mendelssohn, Felix, 68–69
Ménéstrel, Le (journal), 71

Minimalism, 157
Minstrel, 15, 39, 47
Minus One, 141, 142
Mitterrand, François, xii
Mode: in theory of harmony, 151
Moissonneurs, Les (Robert), 97
Molder, 40, 80, 118–19, 128, 129; composer as, 37
Molière, Jean-Baptiste, 49
Money: and music, 3–4, 22, 23, 31, 33, 36–45, 47, 57–62, 124–25, 135, 136, 154
Monody, 154
Monopolization: of music, 8, 51, 52–53, 55, 73; of noise, 27, 124; of violence (of death), 28, 30
Montesquieu, Charles-Louis de Secondat, Baron de, 13
Monteverdi, Claudio, 28, 35, 155
Morrison, Jim, 105, 125
Moses, 87
Mossolov, Alexander, 136
Mozart, Constance, 70
Mozart, Wolfgang Amadeus, 5, 35, 43, 50, 70, 100, 111
Music: as autonomous activity, ix, 3, 47, 51; before capital, 12–18, 46; as commodity, 5, 24, 37, 72, 81, 90, 103, 121, 126, 128–30, 147, 157; as creative of order, 25, 26, 29, 30, 31, 33–34, 57, 60–62, 143; as exchange, 4, 68, 74; history of, vii, 10, 13, 19, 150, 152, 156; Italian vs. French, 155; as mirror of reality, 4, 5, 6, 9–10, 88; non-Western, 154; as object of consumption, 8, 32, 36, 100, 110; as prophetic, xi, xiv, 4, 5, 11–12, 19, 30, 57–59, 65, 83, 126, 130, 135, 144, 154; as sign, 5, 24; as silencing, 19, 22, 47, 88, 105, 111, 118, 120–124, 149–50; as simulacrum of decentralization of power, 114; as simulacrum of representation, 119; as simulacrum of sacrifice, 4, 24, 27–31, 36, 48, 59, 60, 89, 100, 142; social utility of, 24, 57; subversive role of, 4, 6, 8, 11, 13–14, 20, 34, 73, 122, 132, 140–41, 143; as syntax, 35, 36; and use-value, 25, 32, 59. *See also* Aleatory music; Atonal music; Classical music; Court music, Electronic music; Harmony; Liturgical music; Networks of music; New Wave; Polyphony; Popular music; Religious

music; Rhythm and blues; Rock music;
Romantic music; "Serious" music; Written music; Value
Musician: before capital, 11, 12–18, 72; as
domestic, 15–18, 46, 47–50; normalization of, 62–64; as productive worker,
37–41; as rentier, 40, 97, 99; role of in
composition, 140–41; role of in French
Revolution, 55–56; role of in repetition,
47, 57, 84, 96, 106, 112–13, 115, 116,
118; role of in representation, 52–54,
68–77; as subversive, 11–12; as vagabond, 14–15; as wage earner, 38, 80
Musician-priests, 12
Musicians' associations, 78, 79, 80, 84, 89,
99, 105, 138
Musicology, 10, 19, 27, 112, 136, 149,
150–52, 153, 154, 156
Muzak, 8, 111–12, 152
Myth, 28–29

Napoleon III, 78
National Guard (France), 55
National Institute of Music (France), 55–56
Nationalization: of music, 54–57
Nazism, 8
Neo-Hegelianism, 151–52
Neo-tonality, 157
Networks (orders, periods) of music, 10,
19–20, 31–33, 39, 146; decline of, 135,
143; simultaneity of, 42–45; transitions
between, 146
New Wave, 156–58
Newscasts, 108
Nietzsche, Friedrich, 6, 9
Night clubs, 109, 118, 119
Nikish, Artur, 92
Ninth Symphony (Beethoven), 100
Noise: control of, 8, 122–24; and difference,
19, 22, 45; as disorder, 33, 35; entry into
music of, 10, 136–37; in New Wave
music, 157; ordering of, 4, 6, 23, 25, 26,
33–34, 60, 154; and power, 6–9, 27;
silenced by music, 111, 120–21; as
simulacrum of murder, 26, 143
Nono, Luigi, 116
Normalization, 120, 124, 125, 131; crisis of,
85; ideological, 119; of musicians, 62–64;
of speech, 88
Nosologica methodica (Sauvages), 27

Notation: musical, viii, 61
Nouveau Monde, 51
Nouvelles chansons (Boukay), 76

Oboe, 64
Oliveros, Pauline, 158
O'Neill, David, 112
Opera, 17, 154–55, 156, 157
Opera (Paris), 56, 60
Opéra-Comique (Paris), 50
Orchestra, 10, 12, 13, 64, 65–67, 152, 157
Orchestra leader, 64, 66–67, 94, 114, 152
Orders of music. See Networks of music
Original Dixieland Jazz Band, 104
Orontea (Cesti), 155
Overdetermination, viii

Pacific 231 (Honegger), 136
Paganini, Niccolò, 71
Painting, 97, 137
Paleophone, 90–91
Parizot, V., 77
Parliamentary democracy, music in, 8–9, 156
Pas d'acier (Prokofiev), 136
Patents, 129
Patronage, 49, 69
Paulus (singer), 77
Performance: copyright protection for, 79;
development of market for, 78–79; in
repetition, 106; role of in relation to
phonograph record, 84, 85; solo, 67, 154,
156
Performance art, 157, 158
Performers (interpreters): classical, 68–72;
opposition to phonograph record, 96; as
productive workers, 39; as rentiers,
99–100; role of in repetition, 105–7, 114,
118, 152
Peri, Jacopo, 155
Periods of music. See Networks of music
Persée (Lully), 48–49
Philidor (François-André Danican), 51
Phonogram, The, 93
Phonograph, 90, 91–96, 97
Phonograph record, 31, 32, 41, 68, 85,
96–97, 98, 103, 121, 141; 33-RPM (long-
playing), 95, 102, 105; 45-RPM, 104;
78-RPM, 95, 96, 104; industry, 84,
92–99, 108, 152, 157
Phonotaugraph, 90–91

Piano, 35, 69, 119
Picasso, Pablo, 81, 137
Piccini, Niccolo, 63
Pied Piper of Hamelin, 29
Piracy (plagiarism): of music, 54, 57, 98, 164 n68
Plato, 33-34
Polin (comic), 76
Polyphony, viii, 9, 10, 15, 17, 35, 52, 151, 154
Popular (light) music: Attali's interest in, xi, 153; and copyright, 52, 72; and opera, 155; as opposed to "serious" music, 117, 140; relation to court music, 17; in repetition, 102-11, 118, 119-20, 137; "taming" of, 73; and violence, 130, 140, 153, 155; and youth, 10
Pottier (songwriter), 76
Power: accommodation of music to, 119; and background music, 111; and concert hall, 117-18; gain of, by bourgeoisie, 50; loss of in aleatory music, 114-15; music as origin of, 6; music used for, 20, 24, 28, 49; and musician, 11-13, 17, 112, 116-17, 118-19; and orchestra, 65-67; and political economics, 26; recording sound as, 86, 87, 92; and repetition, 90, 100-101, 121-22, 125, 132, 140
Pricing, 41, 59, 100, 101, 107-8, 129, 145
Prima prattica, 154
Printing, 35
Production. *See* Mass production
Productive labor, 37-41
Prokofiev, Sergey Sergeyev, 136
Proliferation. *See* Crisis, of proliferation
Psychoanalysis, 143
Public Health Committee (France), 56
Publicity. *See* Marketing
Publishing: of music, 39, 52-54, 56, 69, 72-73, 74, 96, 99, 144, 166 n116
Puchberg, Michael, 70
Pythagoras, 13, 59, 113

Quantz, Johann Joachim, 16-17
Quartet in C# Minor (Beethoven), 152

Radcliffe-Brown, A. R., 30
Radio, 41, 95, 99, 100, 102, 111; and copyright, 80, 84, 96-97; FM, 105; illegal stations, 131; newscasts on, 108

Rameau, Jean Philippe, 150-51
Rationalization: of music, viii, xi
Ravel, Maurice, 28
Record. *See* Phonograph record
Recording, 35, 85, 87-96, 144; electrical, 95, 96; of images, 144; of music, 32, 141; of speech, 86, 92; tape, 99-100
Recording industry, 84, 92-99, 108, 152, 157
Redding, Otis, 102
Reinach, Th., 99
Religion, 30, 61-62
Religious music, 14, 15
Rent, 40, 81, 97, 99
Repertory, 68-69
Repetition, 20, 33, 40, 42, 134, 137, 141, 144, 145, 147, 152, 156, 158; defined, 41; French term for, 163 n43; rise of, 81-86; role of technicians in, 105-6
Representation, 20, 31-32, 33, 40, 43, 134, 141, 143, 144, 145, 147, 154, 156; decline of, 83, 88-89; defined, 41; French term for, 163 n43; political, 64
Revault d'Allones, O., 140
Revue musicale, La, 73, 78
Rhythm and blues, 104
Richard the Lionhearted, 14-15
Richepin, Jean, 76
Richter (composer), 43
Ries, Ferdinand, 71
Robert Léopold, 97
Rock music, 10, 27-28, 103-5, 156
Rockwell, John, 158
Rohan Chabot, count of, 49
Rolling Stones, 137
Romantic music, 34
Rosetti, Antonio, 43
Rossini, Gioacchino, 164 n68
Round dance, 23, 45
Rousseau, Jean-Jacques, 25, 60, 161 n3, 165 n73
Royal Academy of Music (France), 51, 52
Royal School of Music (France), 56
Royalties, 39, 40, 77-78, 79-80, 84, 98-101, 104, 129, 139, 165 n98
Russian formalism, ix
Russolo, Luigi, 10, 136

SACEM, 78, 79, 80, 84, 89
Sacrifice, 4, 12, 20, 89, 120, 124, 125, 134, 141, 143, 145, 147

Saint Matthew Passion, 68–69
Salis, Rodolphe, 76
Sandow, Gregory, 158
Sant' Onofrio, Conservatorio di, 63
Sapir, Edward, ix
Satie, Erik, 81
"Satisfaction" (Rolling Stones), 137
Saussure, Ferdinand de, vii, 25
Sauvages de la Croix, François Boissier de, 27
Scale: musical, 34, 52, 60
Scarlatti, Alessandro, 155
Schaeffer, Pierre, 9
Schenker, Heinrich, 151–52
Schoenberg, Arnold, x, xi, 43, 81, 82, 153
Schubert, Franz, 35
Schwartzkopf, Elisabeth, 106
Science, 60–61, 65, 89, 91, 113, 150, 151
Scientism, 113
Score: musical, 31, 37, 39–40, 51, 52, 54, 66, 98, 114, 144, 147
Scott, Léon, 90–91
Self-management, 5, 114, 134, 137
Serial music, x, 9, 34, 35, 157
Serinette (bird organ), 73
"Serious" ("learned") music, 102, 112, 113, 115, 117, 120, 140, 157
Serres, Michel, 6, 9, 61
Sex Pistols, 157
Shakespeare, William, 62
Shankar, Ravi, 102
Shepp, Archie, 138, 139
Sibelius, Jean, x
Siège de Corinthe (Rossini), 164 n68
Signs: as commodities, 127, 130; economy of, 37; exchange of, 4, 53, 88; music as, 5, 24; stockpiling of, 131
Silence (Cage), 137
Silencing. *See* Music, as silencing
Socialism, 10–11, 44, 107, 131
Socialist Party (France), xii–xiii
Solo performance, 67, 154, 156
Sonata, 154
Sonata no. 6 (Beethoven), piano, 35
Soul music, 10, 104
Sparta, 13
Spectacle, 88
Spengler, Oswald, vii, x, 18
Spohr, Ludwig, 71

Ssu-ma Ch'ien, 29
Stamitz, Johann, 43
Star: classical, 68–72; popular, 72–77
Star system, 4, 77, 88
State, 42, 49, 55, 56, 146
Stege, Fritz, 8
Stile rappresentativo, 154
Stockhausen, Karlheinz, 113, 131, 142
Stockpiling: of code, 30, 88; of death, 125; and economics, 124, 128, 130, 132, 134, 143, 145; of images, 144; of memory, 87; of music, 32, 41, 84, 95, 101, 111, 141, 153; points, 31; and sales, 109; and valorization, 69
Stravinsky, Igor, 28
"Street Fighting Man" (Rolling Stones), 147
Street hawkers, 72
Street singers, 73–74
Style, 154, 157; elimination of, 115, 116
Sublimation, 23, 26
Suicide, 126, 143
Suicide Motels, 126
Sully-Prudhomme (René-François-Armand Prudhomme), 76
Superstructure: music as part of, vii–viii, xi
Supply. *See* Music, as commodity
Surplus-value, 38, 40, 41–42
Swieten, Baron van, 70
Syndicat des Auteurs, Compositeurs, et Editeurs de Musique (SACEM), 78, 79, 80, 84, 89

Tainter, Charles Sumner, 94
Tamagno, Francesco, 98
Tangerine Dream, 18
Technicians, in repetitive music, 105–6
Technocracy, 112, 117
Technology, viii, xiii, 35, 85, 87, 90, 94, 100, 106, 115, 144, 147
Telephony, 92
Television, 41, 80, 144
Temperament, equal, 61
"Temps des cerises" (popular song), 76
Theresa (Theresa Titiens), 76
Thornton, Clifford, 139
Thoroughbass, 61
Time, 9, 147, 151; stockpiling of, 145. *See also* Use-time
Time Cycle (Foss), 116

Toeschi, Carl Joseph, 43
Tonality, 9, 34, 35, 43, 46, 82, 83, 150–51, 156; neo-, 157
Totalitarianism, 7–8, 55, 153
Touring Club of France, 123
Traité de l'harmonie (Rameau), 151
Tristan und Isolde (Wagner), 143
Troilus and Cressida (Shakespeare), 62
Troubadours, 15
Twelve-tone music. See Serial music

Universal Copyright Convention, 79
Universal music, 113
Use-time, 45, 101, 124–25, 126, 129, 132, 135
Use-value, 9, 25, 32, 41, 42, 59, 84, 96, 99, 101, 103, 107–8, 116, 130, 131, 142
Utopia, xiii, xiv, 133, 145, 147

Value: creation by music, 36–39, 41; of music, ix–x, 51, 58, 78, 106, 107–8, 116, 141. See also Exchange-value; Labor-value; Use-value
Video-disk, 129
Vienna, 81
Viens poupoule, 75
Villeteau, 60
Violence: and chance, 90; channeled by music, 12, 13, 21, 25–31, 36, 46, 57, 59,

89, 110, 150, 152; and death, 125–26; and noise, 20, 26–27, 28, 45, 152; return to, 121, 130–31; ruptured channelization of, 83, 120, 144, 155
Violin, 73
Volapuk, 92, 116

Wagenseil, Georg Christoph, 43
Wagner, Richard, 11–12, 39, 43, 71, 81–82
Weber, Max, viii–ix, x, xi
Webern, Anton von, 28, 35
Whiteman, Paul, 104
Whorf, Benjamin, ix
Women, in composition, 157, 158
"Work" of music: legal definition of, 54, 56, 77
World of Yesterday, The (Zweig), 82
Wranitzky, Paul, 50
Written music, 14, 15, 17

Xenakis, Iannis, 113

Young, Lamont, 35
Youth culture, 10, 104, 105, 109–11

Zhdanov, Andrei Aleksandrovich, 7–8
"Zu den drein Schwanen" (inn), 50
Zweig, Stefan, 82

Jacques Attali is the author of numerous books, including *Millennium: Winners and Losers in the Coming World Order.*

Brian Massumi holds a fellowship with the Australian Research Council in the Department of English at the University of Queensland.

Fredric Jameson is William A. Lane, Jr., Professor of Comparative Literature at Duke University.

Susan McClary is professor of musicology at the University of California, Los Angeles.